"We are all . . . children of this universe. Not just Earth, or Mars, or this System, but the whole grand fireworks. And if we are interested in Mars at all, it is only because we wonder over our past and worry terribly about our possible future."

— Ray Bradbury, 'Mars and the Mind of Man,' 1973

Cover Art: An artist's concept depicting one of many potential Mars exploration strategies. In this approach, the strengths of combining a central habitat with small pressurized rovers that could extend the exploration range of the crew from the outpost are assessed. Rawlings 2007.

NASA/SP–2009–566

Human Exploration of Mars Design Reference Architecture 5.0

Mars Architecture Steering Group
NASA Headquarters

Bret G. Drake, editor
NASA Johnson Space Center, Houston, Texas

July 2009

ACKNOWLEDGEMENTS

The individuals listed in the appendix assisted in the generation of the concepts as well as the descriptions, images, and data described in this report. Specific contributions to this document were provided by Dave Beaty, Stan Borowski, Bob Cataldo, John Charles, Cassie Conley, Doug Craig, Bret Drake, John Elliot, Chad Edwards, Walt Engelund, Dean Eppler, Stewart Feldman, Jim Garvin, Steve Hoffman, Jeff Jones, Frank Jordan, Sheri Klug, Joel Levine, Jack Mulqueen, Gary Noreen, Hoppy Price, Shawn Quinn, Jerry Sanders, Jim Schier, Lisa Simonsen, George Tahu, and Abhi Tripathi.

Available from:

NASA Center for AeroSpace Information
7115 Standard Drive
Hanover, MD 21076-1320
Phone: 301-621-0390 or
Fax: 301-621-0134

National Technical Information Service
5285 Port Royal Road
Springfield, VA 22161
703-605-6000

SUBMITTED UNDER THE GUIDANCE OF THE MARS ARCHITECTURE STEERING GROUP

The human exploration of Mars would be a complex undertaking. It is an enterprise that would confirm the potential for humans to leave our home planet and make our way deep outward into the cosmos. Though just a small step on a cosmic scale, it would be a significant one for humans, because it would require leaving Earth on a long mission with very limited return capability. The strategy and implementation concepts that are described in this report should not be viewed as constituting a formal plan for the human exploration of Mars. Instead, this report provides a vision of a potential approach for human Mars exploration that is based on best estimates of what we know. This is the latest in a series of Mars reference missions that are used by NASA to provide a common framework for future planning of systems concepts, technology development, and operational testing. In addition, this architecture description provides a reference for integration between multiple agency efforts including Mars robotic missions, research that is conducted on the International Space Station, as well as future lunar exploration missions and systems. The strategy outlined in this report was developed from the 2007 Mars architecture study. The Mars Architecture Working Group (MAWG) was comprised of agency-wide representatives from the Exploration Systems Mission Directorate (ESMD), Science Mission Directorate (SMD), Aeronautics Research Mission Directorate (ARMD), and Space Operations Mission Directorate (SOMD). A Joint Steering Group of NASA Agency senior leadership was established in January 2007 to providing insight, guidance, and ultimately, concurrence of recommendations made by the MAWG. This strategy will be updated and revised as we learn more about Mars as well as the systems and technologies that are necessary to conduct human exploration missions beyond low-Earth orbit.

Geoffrey L. Yoder
Director, Constellation Systems Division
Exploration Systems Mission Directorate

Doug McCuistion
Director, Mars Exploration Program
Science Mission Directorate

Thomas B. Irvine
Deputy Associate Administrator
Aeronautics Research Mission Directorate

David P. Radzanowski
Deputy Associate Administrator
Space Operations Mission Directorate

Contents

1	Introduction	1
2	Design Reference Architecture 5.0 Summary	2
2.1	*Surface Reference Mission*	5
2.2	*Communication and Navigation*	8
3	Goals and Objectives	10
3.1	*Mars Planetary Science Objectives (Goals I–III)*	10
3.1.1	Taking advantage of the unique attributes of humans in scientific exploration	10
3.1.2	Scientific objectives for Mars: present and future	11
3.1.3	Geology scientific objectives for the initial human exploration of Mars	13
3.1.4	Geophysics scientific objectives for the initial human exploration of Mars	13
3.1.5	Atmosphere/climate scientific objectives for the initial human exploration of Mars	14
3.1.6	Biology/life scientific objectives for the initial human exploration of Mars	16
3.1.7	The search for extant life	16
3.1.8	The search for past life	17
3.2	*Objectives Related to Preparation for Sustained Human Presence (Goal IV+)*	17
3.2.1	Sustainability-related objectives for the initial human missions to Mars	18
3.3	*Objectives Related to Other Classes of Science (Goal V)*	18
3.3.1	Heliophysics of Mars' environment	19
3.3.2	Space weather	19
3.3.3	Laser ranging for astrophysics	19
3.4	*Goals and Objectives Summary Implications*	19
4	Transportation Systems	21
4.1	*Interplanetary Trajectory and Mission Analysis*	21
4.2	*Heavy-Lift Launch Vehicle*	23
4.3	*In-Space Transportation: Nuclear Thermal Rocket Reference*	25
4.4	*In-Space Transportation: Chemical/Aerocapture Option*	27
4.4.1	Trans-Mars injection module	28
4.4.2	Mars orbit insertion modules (cargo and crew missions)	28
4.4.3	Trans-Earth injection module	28
4.4.4	Low-Earth orbit assembly reboost module	28
4.5	*Launch Processing*	29
4.6	*Crew Exploration Vehicle/Earth Return Vehicle*	30
4.7	*Mars Entry, Descent, and Landing*	31
4.8	*Mars Transit Habitat*	34
5	Surface Systems	36
5.1	*Surface Habitation Systems*	37
5.2	*Surface Mobility Systems*	38
5.2.1	Surface transportation: unpressurized and pressurized rovers	39
5.3	*In-Situ Resource Utilization*	40
5.4	*Surface Power Systems*	41
5.4.1	Stationary power	41
5.4.2	Mobile power	42
6	Architectural Assessments	44
6.1	*Figures of Merit*	45

6.1.1	Safety and mission success	45
6.1.2	Effectiveness	46
6.1.3	Affordability	46
6.2	*Decision 1: Mission Type*	47
6.2.1	Mission-class scientific position	48
6.2.2	Crew health and performance assessment	49
6.2.3	Risk and cost assessments	49
6.2.4	Mission type recommendation	50
6.3	*Decision 2: All-up vs. Pre-deploy Cargo*	51
6.3.1	Pre-deploy option	51
6.3.2	All-up mission option	52
6.3.3	Cargo mission mode recommendation	52
6.4	*Decision 3: Aerocapture vs. Propulsive Mars Orbit Capture of Cargo*	53
6.4.1	Overview of aerocapture	53
6.4.2	System-level trades	54
6.4.3	Aerocapture for the reference payload	54
6.4.4	Aerocapture vs. propulsive capture comparison	55
6.4.5	Aerocapture recommendation	55
6.5	*Decision 4: In-Situ Resource Utilization for Mars Ascent*	56
6.5.1	In-situ resource utilization operational concept	57
6.5.2	In-situ resource utilization trades performed	57
6.5.3	In-situ resource utilization trade study results	58
6.5.4	In-situ resource utilization recommendation	59
6.6	*Decision 5: Mars Surface Power*	60
6.6.1	In-situ resource utilization phase	60
6.6.2	Crewed phase	60
6.6.3	Solar power system concept	60
6.6.4	Fission Surface Power System	61
6.6.5	Large-scale Radioisotope Power System	61
6.6.6	Implementation considerations	61
6.6.7	Surface power system recommendation	62
7	Key Challenges	64
7.1	Human Health and Performance	64
7.1.1	Radiation protection	65
7.1.2	Reduce-gravity countermeasures	65
7.1.3	Medical care	66
7.1.4	Supporting human life	66
7.1.5	Behavior, performance, and human factors	67
7.2	Space Transportation	68
7.2.1	Earth-to-orbit transportation	68
7.2.2	Advanced chemical propulsion	68
7.2.3	Nuclear thermal propulsion	68
7.2.4	Entry, descent, and landing	69
7.3	Surface Systems	69
7.3.1	Advanced habitation systems	69
7.3.2	Extravehicular activity and surface mobility	70
7.3.3	Subsurface access	70
7.3.4	Nuclear power generation	71
7.3.5	Solar/regenerative fuel cell power systems	71
7.3.6	Isotope power systems	72
7.4	*Cross-Cutting Systems and Miscellaneous Needs*	72
7.4.1	In-situ resource utilization	72
7.4.2	Cryogenic fluid management	73
7.4.3	Communication and navigation	74

7.4.4	Supportability and maintainability	74
7.5	*Risk Mitigation Strategies*	75
7.5.1	International Space Station and near-Earth missions	75
7.5.2	Lunar missions	75
7.5.3	Mars robotic missions	76
8	Public Participation	78
8.1	*Guiding Principles for Public Engagement*	79
8.2	*Public Engagement Strategy*	79
Appendix A: Membership		81
Appendix B: References		83

FIGURES

2-1	Mission sequence timelines	3
2-2	Mars Design Reference Architecture 5.0 mission sequence summary (NTR reference)	5
2-3	Notional view of the "Commuter" surface architecture	6
2-4	Notional surface exploration traverses	6
2-5	A notional crew surface exploration timeline	7
2-6	Notional operational scenario near special regions	8
4-1	Cargo and crew trajectories for the example 2037 mission	22
4-2	Trajectory delta-V summary for the reference conjunction class missions	23
4-3	Reference Ares V launch vehicle	24
4-4	Launch vehicle payload performance (45.0.2 configuration)	24
4-5	Crewed and cargo NTR design concepts	26
4-6	Chemical/aerobrake cargo and crewed MTV concepts	27
4-7	Mars aeroshell and payload lifting operations in the VAB	29
4-8	Orion crew exploration vehicle	30
4-9	Orion TPS assessment results	31
4-10	Distributed heating rates for the aerocapture and entry phases	33
5-1	Mars habitats draw from lunar architecture options	38
5-2	Power estimates for solar and nuclear power options	42
6-1	Mars Architecture Study top-level trade tree	44
6-2	Comparison of (a) Opposition-class and (b) Conjunction-class mission profiles	48
6-3	Example mission timeline comparison for Conjunction-class missions	52
6-4	Aerocapture flight profile	53
6-5	ISRU trade tree	58
6-6	Mass, power, and volume of ISRU strategies	59
6-7	Surface power concepts	61
8-1	Sample public engagement pathways	80

TABLES

3-1	Summary of Objectives for the Initial Program of Human Missions to Mars	11
3-2	Goal IV+ Objective Decomposition	17
4-1	Reference NTR Manifest Summary	27
4-2	Reference Chemical/Aerocapture Manifest Summary	29
4-3	EDL System Characteristics	33
4-4	Transit Habitat Mass Summary	34
5-1	Mass Summary for the "Commuter" Surface Scenario	37
5-2	In-Situ Resource Utilization System Mass	40
5-3	Small Pressurized Rover Power Options	43
6-1	Summary of Human Health Mission Type Considerations	50
6-2	Mission Type Recommendation Summary	51
6-3	Cargo Deployment Recommendation Summary	53
6-4	Aerocapture Recommendation Summary	56
6-5	ISRU Recommendation Summary	59
6-6	Power Recommendation Summary	62

ACRONYMS AND ABBREVIATIONS

AEDL	aerocapture and entry, descent, and landing
AG	artificial gravity
ALARA	as low as reasonably achievable
ALSEP	Apollo lunar surface experiments package
Ar	argon
AR&D	automated rendezvous and docking
ARMD	Aeronautics Research Mission Directorate
ATO	abort-to-orbit
AU	astronomical unit
BNTR	bimodal nuclear thermal rocket
CEV	crew exploration vehicle
CFD	computational fluid dynamics
CFM	cryogenic fluid management
CH_4	methane
CHP	crew health and performance
CO	carbon monoxide
CO_2	carbon dioxide
COSPAR	Committee on Space Research
CxP	Constellation Program
DAV	descent/ascent vehicle
DIPS	Dynamic Isotope Power System
DM	docking module
DRA	design reference architecture
DRM	design reference mission
DSN	Deep Space Network
ECLSS	Environmental Control and Life Support System
EDL	entry, descent, and landing
EDS	Earth departure stage
EI	entry interface
EM	electromagnetic
ERV	Earth return vehicle
ESAS	Exploration Systems Architecture Study
ESMD	Exploration Systems Mission Directorate
ETDP	Exploration Technology Development Program
EVA	extravehicular activity
Fe	iron
FOM	figure of merit
FRSI	felt reusable surface insulation
FSP	fission surface power
FSPS	Fission Surface Power System
GaAs/Ge	gallium arsenide/Germanium
GCM	global circulation model

GN&C	guidance, navigation, and control
GPHS	general-purpose heat source
H_2	hydrogen
H_2O	water
H_2O_2	hydrogen peroxide
H_2S	hydrogen sulfide
HCN	hydrogen cyanide
He	helium
HEM-SAG	Human Exploration of Mars Science Analysis Group
HGA	high-gain antenna
HLLV	heavy-lift launch vehicle
HLR	human lunar return
HTPB	hydroxy-terminated polybutadiene
IMLEO	initial mass in low-Earth orbit
Isp	specific impulse
ISRU	in-situ resource utilization
ISS	International Space Station
JSC	Johnson Space Center
klbf	pounds force
KSC	Kennedy Space Center
L/D	lift-to-drag ratio
LAT	Lunar Architecture Team
LCH_4	liquid methane
LEO	low-Earth orbit
LH_2	liquid hydrogen
LOX	liquid oxygen
LRV	lunar rover vehicle
MAWG	Mars Architecture Working Group
MEA	membrane-electrode-assembly
MEL	master equipment list
MEP	Mars Exploration Program
MEPAG	Mars Exploration Program Analysis Group
MER	Mars exploration rover
MLI	multilayer insulation
MOI	Mars orbit insertion
MOLA	Mars orbiter laser altimeter
MRO	Mars Reconnaissance Orbiter
MSL	Mars Science Laboratory
MSR	Mars Sample Return
MTV	Mars transfer vehicle
^{237}Np	Neptunium
N_2	nitrogen
NCRP	National Council on Radiation Protection
NERVA	Nuclear Engine for Rocket Vehicle Applications
NPF	Nuclear Processing Facility
NRC	National Research Council
NTP	nuclear thermal propulsion
NTR	nuclear thermal rocket

O/F	outfitting
O_2	oxygen
OLUID	on-line user identification
OSF	Offline Stacking Facility
PBL	planetary boundary layer
PEMFC	proton exchange membrane fuel cell
PICA	phenolic impregnated carbon ablator
PLSS	Portable Life Support System
psi	pounds per square inch
^{238}Pu	plutonium
PV	photovoltaic
PVA	photovoltaic array
RCS	Reaction Control System
RFC	regenerative fuel cell
RPS	Radioisotope Power System
RSRB	reusable solid rocket booster
RTG	radioisotope thermoelectric generator
RWGS	reverse water gas shift
SHAB	surface habitat
SLA	super lightweight ablator
SM	service module
SMD	Science Mission Directorate
SO_2	sulfur dioxide
SOCE	solid oxide carbon dioxide electrolyzer
SOMD	Space Operations Mission Directorate
SPE	solar particle event
SRB	solid rocket booster
t	ton (metric)
TDRSS	Tracking and Data Relay Satellite System
TEI	trans-Earth injection
TMI	trans-Mars injection
TPS	Thermal Protection System
TRL	technology readiness level
TT&C	telemetry, tracking, and communications
VAB	Vehicle Assembly Building
ZBO	zero boil-off
ZBR	Zone of Minimum Biological Risk

1 INTRODUCTION

The NASA Authorization Act of 2005 articulated a new strategy for the nation's space program by specifically stating that "The Administrator shall establish a program to develop a sustained human presence on the Moon, including a robust precursor program, to promote exploration, science, commerce, and United States preeminence in space, and as a stepping-stone to future exploration of Mars and other destinations." This vision calls for a progressive expansion of human capabilities beyond low-Earth orbit (LEO), seeking to answer profound scientific and philosophical questions while responding to discoveries along the way. In addition, the strategy calls for developing the revolutionary new technologies and capabilities that are required for the future human exploration of the solar system. This strategy represents a bold new step.

2020 Vision – Humans bring experience, ingenuity, and adaptability to enable robust exploration and discovery. Rawlings 1997.

In January 2004, NASA established the Exploration Systems Mission Directorate (ESMD) to lead the development of new exploration systems to accomplish the task of implementing the strategy. To determine the best exploration architecture and strategy to implement these many changes, the Exploration Systems Architecture Study (ESAS) was conducted in 2005. This study provided the top-level architectural foundation and driving requirements for the lunar transportation systems. In 2006 through mid-2007, NASA conducted the Lunar Architecture Team (LAT) series of studies, which was aimed at further definition of the goals and objectives, activities, and systems necessary for conducting the lunar surface portion of the exploration strategy. Whereas the ESAS focused on the transportation system, the lunar architecture assessments concentrated on the activities conducted on the surface.

During execution of the second half of the LAT studies, it was recognized that the lunar definition must be conducted in an environment that considers the most likely follow-on mission, namely the human exploration of Mars. Significant progress was being made in the definition of the lunar architecture and systems, but further refinement and confirmation of how these systems would either be used, or modified, for future exploration capabilities was required. In addition, the Science Mission and Aeronautics Research Mission Directorates were in the process of defining future Mars robotic missions as well as fundamental research activities related to future human exploration missions. Recognizing the need for an updated and unified reference architecture for human exploration of Mars, NASA Headquarters commissioned The Mars Architecture Working Group (MAWG) in January 2007 to develop the Mars Design Reference Architecture 5.0 (DRA 5.0).

The MAWG was comprised of agency-wide representatives from the ESMD, Science Mission Directorate (SMD), Aeronautics Research Mission Directorate (ARMD), and Space Operations Mission Directorate (SOMD). In addition, an Agency Joint Steering Group of senior leadership was established at the beginning of the study to review the primary products that were produced by the MAWG, providing insight, guidance, and, ultimately, concurrence of recommendations made by the team.

The strategy and implementation concepts that are described in this report should not be viewed as constituting a formal plan for the human exploration of Mars. Instead, this report provides a vision of one potential approach to human Mars exploration that is based on best estimates of what we know. This approach is used to provide a common framework for future planning of systems concepts, technology development, and operational testing. In addition, it provides a common reference for integration between multiple agency efforts including Mars robotic missions, research conducted on the International Space Station (ISS), as well as future lunar exploration missions and systems. The strategy outlined in this report will be updated and revised as we learn more about Mars as well as the systems and technologies that are necessary to conduct human exploration missions beyond low Earth orbit.

2 DESIGN REFERENCE ARCHITECTURE 5.0 SUMMARY

The Mars Design Reference Architecture (DRA 5.0) describes the systems and operations that would be used for the first three missions to explore the surface of Mars by humans. These first three missions would occur on three consecutive trajectory opportunities sometime within the next several decades. A minimum three-mission set was chosen for this reference architecture for several reasons:

♦ The development time and cost to achieve the basic capability to carry out a single human Mars mission are of a magnitude that a single mission or even a pair of missions is difficult to justify.
♦ Three consecutive missions would require approximately 10 years to complete; a period of time that is sufficient to achieve basic program goals and acquire a significant amount of knowledge and experience, making this a likely point in time to consider new goals and improved architectures to achieve them.

Commuter – An artist's concept depicting a potential Mars exploration outpost. Rawlings 2007

In addition, these first three human Mars missions are assumed to have been preceded by a sufficient number of test and demonstration missions on Earth, in the ISS, in Earth orbit, on the moon, and at Mars (by robotic precursors) to achieve a level of confidence in the architecture such that the risk to the human crews is considered acceptable. The human exploration of Mars would be a complex undertaking. It is an enterprise that would confirm the potential for humans to leave our home planet and make our way outward into the cosmos. Although just a small step on a cosmic scale, it would be a significant one for humans because it would require leaving Earth on long missions with a constrained return capability. The commitment to launch is a commitment to several years away from Earth, and there is a very narrow window within which return is possible. This is the most radical difference between Mars exploration and previous lunar explorations. Successful implementation of human exploration of Mars will require a thorough and in depth technology development program that is coupled with a rigorous risk mitigation strategy.

For the reference architecture described herein, a crew of six would be sent on each of these missions, and each crew would visit a different location on Mars. The rationale for a crew of this size has been judged to be a reasonable compromise between the skill mix and level of effort for missions of this complexity and duration balanced with the magnitude of the systems and infrastructure needed to support this crew. Visiting three different sites is based on a recommendation from a special committee of the Mars Exploration Program Analysis Group (MEPAG) described in Section 3 of this document. The science and exploration rationale for visiting three different sites recognizes that a planet that is as diverse as Mars is not likely to be adequately explored and understood from the activities that could take place at a single site. However, this three-site assumption does not preclude returning to any of the sites should there be a compelling need to do so.

Each of the three missions would use the conjunction class (long-stay) trajectory option. A portion of each mission's assets would be sent to Mars one opportunity prior to the crew. This, the so-called "pre-deploy" or "split mission" option, would allow a lower energy trajectory to be used for these pre-deployed assets, which allows more useful payload mass to be delivered to Mars for the propellant available. The decision to pre-position some of the mission assets also better accommodates the decision to make part of the ascent propellant at Mars, using the martian atmosphere as the raw material source for this ascent propellant. This use of in-situ resources and the equipment to process these resources into useful commodities results in a net decrease in the total mass that is needed to complete a mission as well as a significant reduction in the size of the landers. A surface nuclear power source, as compared to an equivalent solar power system, was found to be better suited for producing this ascent propellant. This choice was further supported by the fact that this power system would be more than adequate to meet the needs of the human crew members when they arrive, which occurs after all of the necessary propellants have been produced. Splitting the mission elements between pre-deployed cargo and crew vehicles allows the crew to fly on faster, higher-energy trajectories, thus minimizing their exposure to the hazards associated with deep-space inter-planetary travel.

Due to the significant amount of mass required for a human mission to Mars, numerous heavy-lift launches would be required. The reference launch vehicle that would be used is the Ares V lunar cargo launch vehicle. Using the same lunar launch vehicle would greatly improve the overall launch reliability due to the maturity of the launch vehicle by the time the Mars missions commence. Current estimates of the mission manifest indicate that at least seven heavy-lift cargo launches would be required, but the number of launches could be higher, depending on the architecture-wide technology options inserted. This large number of launches necessitates a significant launch campaign that must begin several months prior to the opening of the Mars departure window. The reference strategy that is adopted eliminates on-orbit assembly of the mission elements by segmenting the systems into discrete packages and using automated rendezvous and docking (AR&D) of the major elements in LEO. Launches would occur 30 days apart and would be completed several months before the opening of the Mars departure window to provide a margin for technical delays and other unforeseen problems. This strategy requires that the in-space transportation systems and payloads loiter in LEO for several months prior to departure for Mars. The overall launch and flight sequence for the first two missions is depicted in figure 2-1.

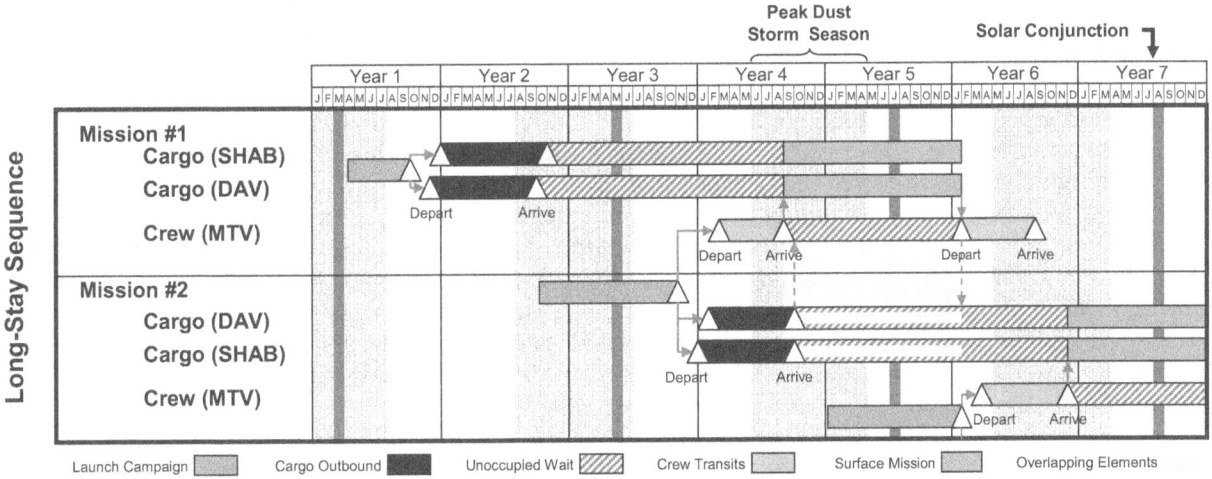

Figure 2-1. Mission sequence timelines.

The first phase of the mission architecture would begin with the pre-deployment of the first two cargo elements, the descent/ascent vehicle (DAV) and the surface habitat (SHAB). These two vehicle sets would be first launched, assembled (via rendezvous and docking), and checked out in LEO. After all of the systems have been verified and are operational, the vehicles loiter in Earth orbit until the Earth-Mars departure window opens when they would be injected into minimum energy transfers from Earth orbit to Mars just over 2 years prior to the launch of the crew. Nuclear thermal rocket (NTR) propulsion was chosen as the reference concept for the in-space propulsion system for both cargo and crew vehicles. Upon arrival at Mars, the vehicles would be captured into a high-Mars orbit. The SHAB would remain in Mars orbit in a semi-dormant mode, waiting for arrival of the crew 2 years later. The DAV would be captured into a temporary Mars orbit from which it would autonomously perform the entry, descent, and landing on the surface of Mars at the desired landing site. After landing, the vehicle would be checked out and its systems verified to be operational. The surface fission reactor would be deployed, and production of the ascent propellant and other commodities that are needed by the crew would be completed before committing to the crew phase of the mission.

A key feature of the long-stay mission architectures is the autonomous deployment of a portion of the surface infrastructure before the crew arrives such as the surface power system. This strategy includes the capability for these infrastructure elements to be unloaded, moved significant distances, and operated for significant periods of time without humans present. In fact, the successful completion of these various activities would be part of the decision criteria for launch of the first crew from Earth.

The second phase of this architecture begins during the next injection opportunity with the launch, assembly, and checkout of the crew Mars transfer vehicle (MTV). The MTV would serve as the interplanetary support vehicle for the crew for a round-trip mission to Mars orbit and back to Earth. Prior to departure of the flight crew, a separate checkout crew may be delivered to the MTV to perform vital systems verification and any necessary repairs prior to departure of the flight crew. After all vehicles and systems, including the Mars DAV (on the surface of Mars), SHAB (in Mars orbit), and the MTV (in LEO) are verified operational, the flight crew would be injected on the appropriate fast-transit trajectory towards Mars. The length of this outbound transfer to Mars is dependent on the mission date, and ranges from 175 to 225 days. Upon arrival at Mars, the crew members perform a rendezvous with the SHAB, which would serve as their transportation leg to the surface of Mars.

Current human health and support data indicate that it may take the crew a few weeks to acclimate to the partial gravity of Mars after landing. After the crew has acclimated, the focus of initial surface activities would focus on transitioning from a "lander mode" to a fully functional surface habitat. This would include performing all remaining setup and checkout that could not be performed prior to landing, as well as transfer of hardware and critical items from the pre-deployed DAV.

The long-stay mission architecture lends itself to a very robust surface exploration strategy. The crew would have approximately 18 months in which to perform the necessary surface exploration. Ample time would be provided to plan and re-plan the surface activities, respond to problems, and readdress the scientific questions posed throughout the mission. The focus during this phase of the mission would be on the primary science and exploration activities that would change over time to accommodate early discoveries. A general outline of crew activities would be established before the launch, but would be updated throughout the mission. This outline would contain detailed activities to ensure initial crew safety, make basic assumptions as to initial science activities, schedule periodic vehicle and system checkouts, and plan for a certain number of sorties. Much of the detailed activity planning while on the surface would be based on initial findings and, therefore, could not be accomplished before landing on Mars. The crew would play a vital role in planning specific activities as derived from more general objectives defined by colleagues on Earth. Alternative approaches for exploring the surface are still under discussion and are expected to be examined further, including maximizing commonality with lunar systems. One of the approaches that most closely follows previous DRAs, referred to as the "Commuter" scenario, was selected as the nominal approach and is described in the next section.

Before committing the crew to Mars ascent and return to Earth, full systems checkout of the ascent vehicle and the MTV would be required. Because both vehicles are critical to crew survival, sufficient time must be provided prior to ascent to verify systems and troubleshoot any anomalies prior to crew use. In addition, the surface habitat would be placed in a dormant mode for potential reuse by future crews. This includes stowing any nonessential hardware, safing critical systems and their backups, and performing general housekeeping duties. Lastly, surface elements, including science instruments, would be placed in an automated operations mode for Earth-based control. The crew would then ascend in the DAV and performs a rendezvous with the waiting MTV. This vehicle would be used to return the crew from Mars, ending with a direct entry at Earth in a modified Orion crew vehicle. The nuclear thermal propulsion (NTP) version of the DRA, also known as a "bat chart", is shown in figure 2-2.

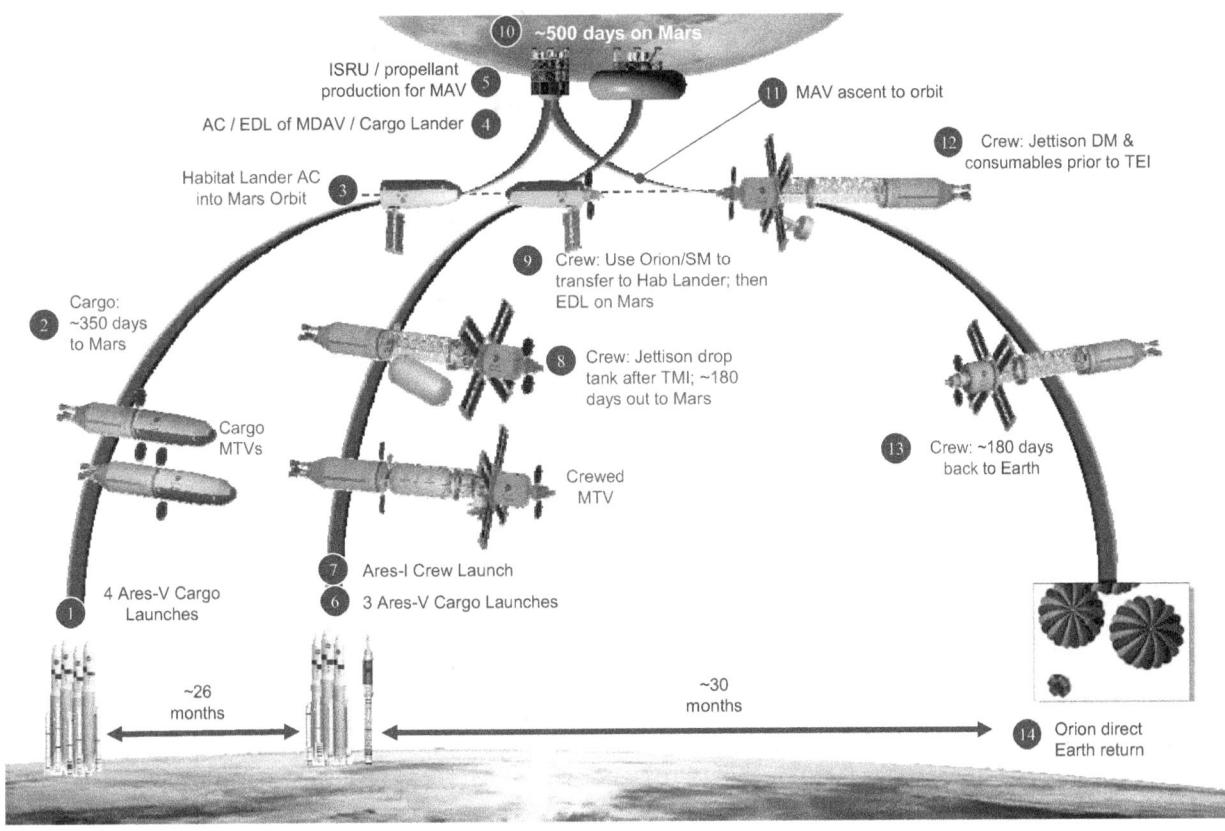

Figure 2-2. Mars Design Reference Architecture 5.0 mission sequence summary (NTR reference).

2.1   *Surface Reference Mission*

Several different surface architectures were assessed during the formulation of the Mars DRA 5.0, each of which emphasized different exploration strategies that were embodied in the combination of duration of in the field, range of exploration reach, and depth of subsurface access. The nominal surface mission scenario adopted for DRA 5.0 is the so-called "Commuter" reference architecture, which would have a centrally located, monolithic habitat (figure 2-3), two small pressurized rovers, and two unpressurized rovers (roughly equivalent to the lunar rover vehicle (LRV) that was used in the Apollo missions to the moon). This combination of habitation and surface mobility capability would allow the mission assets to land in relatively flat and safe locations, yet provides the exploration range that would be necessary to reach nearby regions of greater geologic diversity (figure 2-4). Power for these systems would be supplied by a nuclear power plant that was previously deployed with the DAV and used to make a portion of the ascent propellant. Traverses would be a significant feature of the exploration strategy that would be used in this scenario, but these traverses would be constrained by the capability of the small pressurized rover. In this scenario, these rovers have been assumed to have a modest capability, notionally a crew of two, 100 km total distance before being re-supplied, and 1- to 2-week duration. Thus, on-board habitation capabilities would be minimal in these rovers. However, these rovers are assumed to be nimble enough to place the crew in close proximity to features of interest (i.e., close enough to view from inside the rover or within easy extravehicular activity (EVA) walking distance of the rover). Not all crew members would deploy on a traverse, so there would always be some portion of the crew in residence at the habitat. The pressurized rovers would carry (or tow) equipment that would be capable of drilling to moderate depths – from tens to hundreds of meters – at the terminal end of several traverses.

Candidate surface sites would be chosen based on the best possible data available at the time of the selection, the operational difficulties associated with that site, and the collective merit of the science and exploration questions that could be addressed at the site. Information available for site selection would include remotely gathered data sets plus data from any landed mission(s) in the vicinity plus interpretive analyses based on these data.

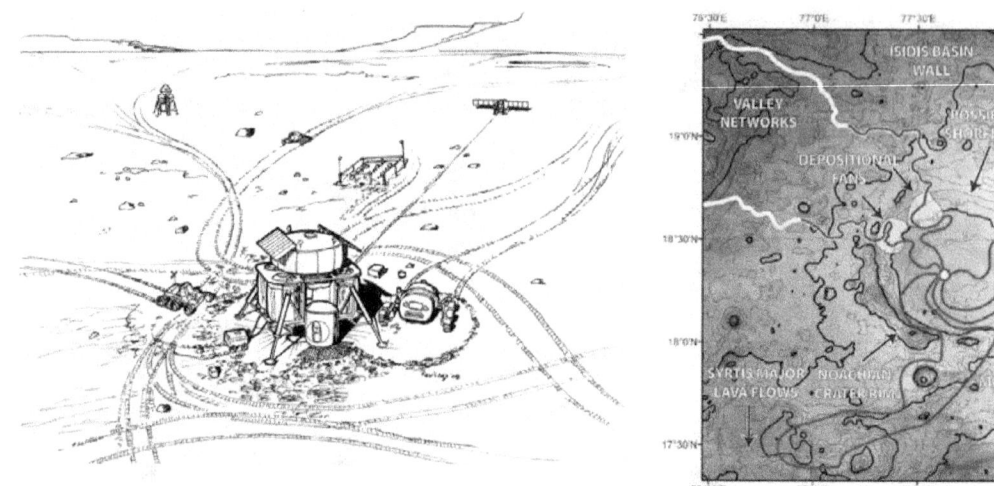

Figure 2-3. Notional view of the "Commuter" surface architecture. **Figure 2-4.** Notional surface exploration traverses.

Figure 2-4 illustrates a notional series of traverses to features of interest at the junction of the Isidis Planatia and Syrtis Major regions. No particular preference is being given to this site; it is included here to illustrate some general features of a human exploration mission and the resulting implications for operations at such a site.

From an operational perspective, this location has a relatively broad, relatively flat, centrally located area where cargo elements could land in relative safety. However, this would place these systems and the crew at large distances from features that are of interest to the crew and the science teams. The scale at the lower right of figure 2-4 indicates that these features of interest are beyond what is currently considered a reasonable walking range for the crew (determined by the distance a crew member could walk during one charge of power and breathing gases in his/her Portable Life Support System (PLSS) – roughly 20 km total. Although sites with much more closely spaced features of interest certainly exist, they are usually found at the expense of a relatively safe landing site.

One feature of interest is not illustrated here – the subsurface. Understanding the vertical structure of the site would also be of interest, indicating that a drilling capability would need to be included for each mission and site. The ability to move a drill from location to location would also be desirable.

The primary habitat would have space and resources allocated for on-board science experiments. The pressurized rovers would carry only the minimal scientific equipment that is deemed essential for field work (in addition to the previously mentioned drill). Samples would be returned to the primary habitat and its on-board laboratory for any extensive analysis.

One approach to accomplishing the desired long traverses would be to use the pressurized rovers (or possibly robotic rovers) to pre-position supplies in caches along the proposed route of travel prior to the "full-duration" traverse. Thus, a typical traverse would begin with the crew (or robotic rovers) traveling out a nominal distance (approximately 15 km, or EVA walk-back distance) and establishing a cache of commodities for life support and power (possibly emergency habitation) before returning to the habitat. Some amount of exploration-related activities may be accomplished during this cache deployment phase, but its primary purpose would be route reconnaissance and cache establishment. The crew would then make another traverse, establishing a second cache a like distance beyond the first cache. This process would continue until all caches in the chain are built up sufficiently for the crew, in the two pressurized rovers, to make the entire round-trip traverse for the time duration needed to accomplish traverse objectives. The amount of time required to set up and retrieve the supply caches would depend on the specific conditions for a traverse. However, the timeline in figure 2-5 illustrates how much could be accomplished if approximately 2 weeks are allocated in which to establish this string of caches and another 2 weeks to retrieve them. In addition, not all traverses would be long enough to require this type of support. A mixture of cache-supported and -unsupported traverses has been illustrated. Finally, some amount of time would be required to repair and restock the pressurized rovers after each traverse, as well as to conduct any local experiments and plan for the next traverse. A notional 2 weeks between short traverses and 4 weeks between long traverses is illustrated in figure 2-5.

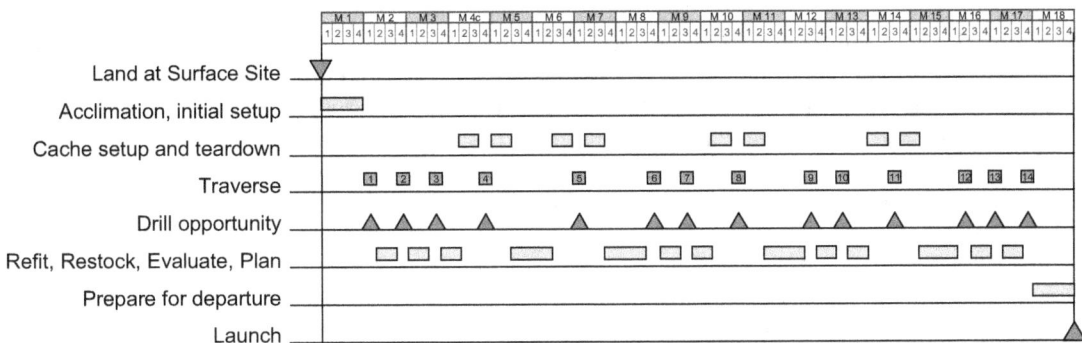

Figure 2-5. A notional crew surface exploration timeline.

A strong motivating factor for the exploration of Mars is the search for extraterrestrial life. However, this search could be permanently compromised if explorers carry Earth life and inadvertently contaminate the martian environment. Additionally, we must guard against the remote possibilities that samples returned from Mars could contain living organisms that might reproduce on Earth and damage some aspect of our biosphere. Preventing both of these eventualities is termed "planetary protection."

International planetary protection policy is maintained by the Committee on Space Research (COSPAR). This policy applies to all planetary exploration carried out by nations that are signatory to the 1967 Outer Space Treaty. Although detailed planetary protection requirements for human missions have not yet been established, guidelines for human missions are currently being developed and incorporated into COSPAR policy in preparation for future human exploration. Although implementation details for human missions may be different from those applied to exclusively robotic missions, the intent of planetary protection policy would remain the same: to avoid harmful contamination of other solar system bodies (forward contamination), and to avoid adverse changes in the environment of the Earth (backward contamination).

In order for humans to explore Mars and return to Earth safely, it will be necessary to identify sites on Mars that are free of hazards to the Earth's biosphere. This is because astronauts on the martian surface inevitably would be exposed to local martian materials such as dust, and the plan is to return the astronauts to Earth at the end of the mission. The astronauts are therefore a potential vector for the transport of martian dust, which must be shown in advance to be sufficiently safe. The Space Studies Board has recommended the designation of Zones of Minimum Biological Risk (ZBRs) that are regions demonstrated to be safe for humans. That is, astronauts would only be allowed in areas that are demonstrated to be safe. For the initial landing site, such testing would probably have been performed as a part of the precursor mission activities, which may include analysis on Earth of returned martian samples, particularly wind-blown dust.

Human beings carry large microbial populations in and on their bodies, and these populations are constantly reproducing. Even with improvements in human support technologies, it is almost certainly impossible for all human-associated processes and mission operations to be conducted within completely closed systems. For this reason, it must be assumed that some microbial contamination of the surrounding martian environment would be inevitable, so human missions should be sent only to sites on Mars where this consequence is acceptable.

The surface of Mars is very cold and dry; in most places, it is too cold or dry to permit the growth and reproduction of Earth organisms. However, certain geological features on Mars may be warmer and wetter, including large recent craters, the mid-latitude gullies and associated "pasted-on" terrain, thermal anomalies (if present), and very young volcanic rocks. In addition, the subsurface of Mars is almost completely unknown, and is even more likely to be habitable by Earth microbes than surface features. Mars Special Regions are currently (2007) defined by COSPAR as "a region within which terrestrial organisms are likely to propagate, OR A region that is interpreted to have a high potential for the existence of extant Martian life forms." Under current understanding this encompasses a small fraction of the surface of the planet, excluding both equatorial and polar latitudes. Special Regions shall only be accessed using sterilized clean equipment, to prevent forward contamination.

Further assessments in the operational scenarios and necessary surface support systems that are needed to explore the surface of Mars productively are required. The strategy that was adopted for the current DRA envisions targeting the human landing site that would be located within an area that is already known to be safe to humans (a zone of minimum biological risk) and in which microbial contamination would be permitted. Astronauts must be protected from contact with untested martian materials, and their health must be monitored to ensure that the results of exposure could be understood. Samples of martian material from places with the potential to serve as habitats for Earth organisms may be collected using sterile sampling rovers or drills (figure 2-6) or via advanced sterilization techniques. Advances in sterilization and cleaning technologies, as well as procedures that could be performed in collaboration with human exploration activities, will be critical to facilitate exploration of astrobiologically interesting sites on Mars. Facilities for handling collected samples under appropriate contamination control would be required to protect science, astronauts, and the Earth.

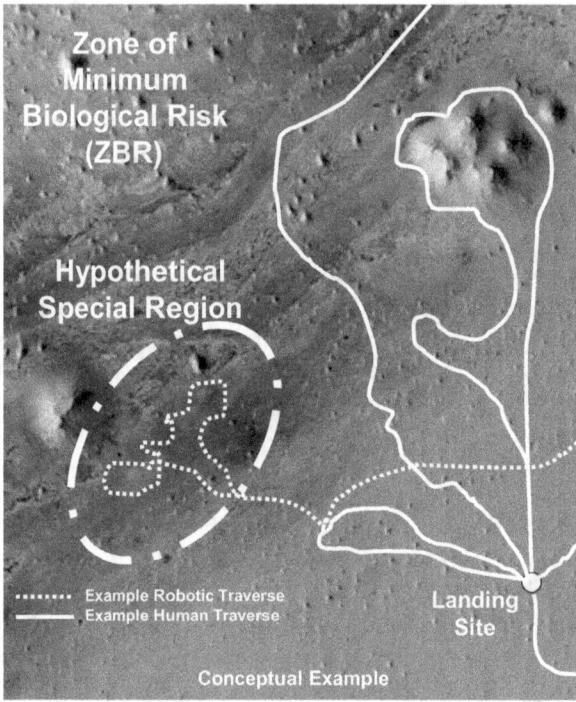

Figure 2-6. Notional operational scenario near special regions.

2.2   *Communication and Navigation*

The overall NASA communications and navigation architecture must support the full scope of Mars exploration, including launch, Earth orbital operations, trans-Mars injection (TMI), Earth-Mars cruise, Mars orbit insertion (MOI), Mars orbital operations, entry/descent/landing, surface operations, Mars ascent, on-orbit rendezvous, TMI, and Earth arrival. Meeting this range of mission phases would require the combined capabilities of the Space Network (for initial near-Earth support), the Deep Space Network (DSN), and dedicated Mars network assets. While detailed communications requirements for Mars exploration are not yet well understood, the feed-forward nature of lunar exploration as a precursor to Mars exploration offers strong motivation for providing comparable communications and navigation capabilities at Mars as would be used at the Moon, thereby supporting similar exploration operations concepts.

The maximum distance between Earth and Mars is roughly 400,000,000 km, 1,000 times the distance between Earth and the moon. Because of the loss of signal strength due to the increased distance, communications from Mars are effectively a million times more challenging than communications from the moon. The large distance to Mars also implies long signal transit times, with round-trip light times of up to 44 minutes. This, too, profoundly affects the basic operations concepts for Mars exploration relative to lunar exploration.

While the lunar communications architecture provides important feed-forward concepts and capabilities, the strategies for Mars would require tailoring to address the much larger distances and resulting communications delays that are involved. Like the lunar architecture, the notional Mars architecture includes a combination of ground stations on Earth and orbital assets at Mars that would provide communications and navigation services to exploration users. However, details of the design, particularly for the long-haul links between Earth and Mars, must be modified. Nevertheless, common solutions would be sought wherever possible, including aspects of the short-range relay links as well as upper layers of the communications protocol stack, above the physical and link layers that are driven by distance, which could remain essentially the same as in the lunar architecture.

From launch through TMI, the NASA Space Network would provide continuously available communications and navigation services via Tracking and Data Relay Satellite System (TDRSS) S-band and Ka-band links. After TMI, support during Earth-Mars cruise would transition to the NASA DSN, with basic telemetry, tracking, and communications (TT&C) functions provided by X-band links and high-rate links supported at the Ka-band or, perhaps, optical communication. A particular challenge during the cruise phase would be ensuring adequate safe mode communications in the event of an anomaly.

Once in the Mars environment (i.e., on final approach, in orbit, in the atmosphere, or on the surface), a user spacecraft would be able to obtain efficient, high-rate communications and tracking services from on-orbit assets. Future trades regarding the orbital asset communication strategy are required. The orbiting assets could be dedicated Mars relay satellites or pre-deployed human mission cargo vehicles, each of which would be outfitted with highly capable direct-to-Earth communications payloads to support high rates on the trunk line back to Earth, allowing individual users to use much smaller, lighter, and lower-power communications systems on the relatively short-range links to the orbital assets. In addition to providing an energy-efficient means for communications between a Mars user and Earth, the Mars relay satellites could also play a key role in supporting communications between spatially separated users at Mars (e.g., between a Mars habitat and an astronaut on a long-range, over-the-horizon excursion); and for users in the immediate line-of-sight vicinity of the Mars habitat, a Mars communications terminal on the surface would provide even more efficient wired and wireless communications options over short-range links.

Options for relay satellites include dedicated telecommunications orbiters as well as the use of other orbital assets with added telecommunications functionality. In the latter category, the MTV, as well as potential cruise stages associated with surface-deployed elements (e.g., habitat, cargo vehicles, assuming deployment from orbit), could provide relay functionality at low cost by incorporating a relay payload. (This strategy has been used successfully for Mars robotic exploration, with remote sensing orbiters in low circular orbit, such as the Mars Global Surveyor, Mars Odyssey, and Mars Reconnaissance Orbiter (MRO), providing valuable relay services to robotic landers such as the Mars exploration rovers (MERs), Phoenix Lander, and Mars Science Laboratory (MSL).) However, telecommunications functionality would typically be limited based on competing demands on the spacecraft orbit. For instance, the MTV orbit strategy would be driven by the demands of MOI, Mars landing site targeting, on-orbit rendezvous, and trans-Earth injection (TEI); these requirements would likely result in an orbit that is not optimal from a telecommunications perspective.

In contrast, a dedicated relay spacecraft could have its orbit optimized for its communications and navigation functions. Various orbits have been considered for dedicated relay assets, including low-altitude circular orbits, mid-altitude circular orbits, elliptical orbits, and areostationary orbits (the Mars equivalent of Earth geostationary orbits). Based on assessments to date, the areostationary orbit option is selected as the most desirable dedicated orbit option, in particular based on its continuous coverage capability. (Nevertheless, future trade studies may wish to explore the telecommunications potential of other planned orbiter vehicles such as the MTV as a low-cost alternative and/or backup to dedicated areostationary orbiter(s).)

Within the vicinity of the Mars SHAB, a Mars communications terminal is envisioned that would provide high-rate, energy-efficient services to users in the immediate surface environment. The Mars communications terminal provides the necessary effective isotropic radiated power and gain/temperature to achieve high-rate links to Mars relay satellites, as well as backup links to the DSN in the event of relay satellite anomalies.

3 GOALS AND OBJECTIVES

After extensive discussion, the MAWG concluded that the goals for the initial human exploration of Mars are best organized under the following taxonomy:

Mars Field Work – Flexible and robust exploration capabilities are key aspects of enabling scientific discoveries. Frassanito 2003.

- *Goals I–III (Planetary Science)*: The traditional goals (MEPAG, 2006) for understanding Mars' potential for life (Goal I), its current and ancient climate (Goal II), and its geology/geophysics (Goal III).
- *Goal IV+ (Preparation for Sustained Human Presence)*: MEPAG (2006) uses the term "Goal IV" to describe preparation for the first human explorers. By definition, this cannot be a goal for the first human missions since, by then, the preparation would have to be complete. However, a goal of the first human missions is to prepare for the subsequent future, which includes sustained human presence on the surface of Mars.
- *Goal V (Ancillary Science)*: This includes all scientific objectives unrelated to Mars, including those related to astrophysics, observations of the sun, Earth, moon, and interplanetary environment. Note that these objectives may be important during the transit phase for missions to and from Mars.

Analysis of Goals I through III was prepared by an analysis team that was sponsored by the MEPAG that went by the name of Human Exploration of Mars Science Analysis Group (HEM-SAG). HEM-SAG produced a substantial white paper (MEPAG HEM-SAG, 2008). Section 2.2 (also see table 3-1) of this report is a summary of that more detailed analysis.

3.1 Mars Planetary Science Objectives (Goals I–III)

Mars is a diverse and complex world. Many of the same processes/mechanisms operate(d) on both Earth and Mars; e.g., early heavy bombardment, impact craters, planetary dipole magnetic field (at least in the early history of Mars), widespread and extensive volcanism, the presence of liquid water (H_2O) on the surface, geochemical cycles, the condensation of atmospheric gases forming polar caps, etc. Mars, like Earth, is a terrestrial planet with very diverse and complex geological features and processes. Like Earth, Mars is also a possible abode for past and/or present life. The geological record suggests that the atmosphere/climate of Mars has changed significantly over its history. Early Mars may have possessed a significantly denser atmosphere that was lost (Jakosky and Phillips, 2001). A denser atmosphere on Mars would have permitted liquid H_2O on its surface. Present-day Mars has a thin (6 millibar pressure at the surface), cold atmosphere that is devoid of any surface liquid H_2O. Why has Mars changed so drastically over its history? How and why has the habitability of Mars changed over its history? Is there a message in the history of Mars to better understand the future of Earth? Did life form on early Mars? Is there evidence of early life in the geological record? Is there life on Mars today?

3.1.1 Taking advantage of the unique attributes of humans in scientific exploration

It is important to consider the unique capabilities that humans bring to the process of exploring Mars. As a result, a common set of human traits emerged that apply to exploration relating to the MEPAG science disciplines, which include geology, geophysics, life, and climate. These characteristics include: speed and efficiency to optimize field work; agility and dexterity to go places that are difficult for robotic access and to exceed currently limited degrees-of-freedom robotic manipulation capabilities; and, most importantly, the innate intelligence, ingenuity, and adaptability to evaluate in real time and improvise to overcome surprises while ensuring that the correct sampling strategy is in place to acquire the appropriate sample set. Real-time evaluation and adaptability especially would be a significant new tool that humans on Mars would bring to surface exploration. There are limitations to the autonomous operations that are possible with current robotic systems, with fundamental limitations to direct commanding from Earth being the time difference imposed by the 6- to 20-minute communications transit time and the small number of daily uplink and downlink communications passes.

Table 3-1. Summary of Objectives for the Initial Program of Human Missions to Mars

Goals I–III	Goal IV+	Goal V
Quantitatively characterize the different components of the **martian geologic system** (at different times in martian geologic history), and understand how these components relate to each other (in three dimensions).	Learn to make effective use of **martian resources** including providing for crew needs and, if possible, power and propulsion consumables.	Ancillary science (heliophysics, astrophysics).
Search for **ancient life** on Mars.	Develop reliable and robust **exploration systems**. Increase the level of self-sufficiency of Mars operations.	
Make significant progress towards the goal of understanding whether or not **martian life** forms have persisted to the present (extant biological processed).	Address **planetary protection** concerns regarding sustained presence.	
Quantitatively understand early Mars **habitability** and early Mars possible **pre-biotic biogeochemical cycles and chemistry**.	Promote the development of **partnerships** (international, commercial, etc.) and sustain **public engagement**.	
Characterize the structure, composition, dynamics, and evolution of the **martian interior** (core to crust).		
Quantitatively understand **martian climate history** with attention to the modern climate/weather system.		

Notes. 1. Not listed in priority order. 2. For Goal V, it was not possible to be specific.

Humans are unique scientific explorers that could obtain previously unobtainable scientific measurements on the surface of Mars. Humans also possess the ability to adapt to new and unexpected situations in new and strange environments. They can make real-time decisions, have strong recognition abilities, and are intelligent. Humans also can perform detailed and precise measurements of the surface, subsurface, and atmosphere while on the surface of Mars using state-of-the-art scientific equipment and instrumentation brought from Earth. The scientific exploration of Mars by humans would presumably be performed as a synergistic partnership between humans and robotic probes – a partnership that is controlled by the human explorers on the surface of Mars (MEPAG HEM-SAG, 2008).

Robotic probes could explore terrains and features not suitable or too risky for human exploration. Under real-time human control, robotic probes could traverse great distances from the human habitat, covering distances/terrain too risky for human exploration; undertake sensitive, delicate sample handling operations; and return rock and dust samples to the habitat for triage and laboratory analyses.

3.1.2 Scientific objectives for Mars: present and future

Our current scientific objectives for the exploration of Mars have been described in detail by MEPAG (2006). However, in planning the scientific objectives of a mission that would be undertaken 20 to 25 years from now, we also need to take into account the additional robotic missions that are likely to be scheduled before the first human mission, and the progress that these robotic missions would make towards achieving these objectives. We need to plan the objectives of a 2030 mission based on our projected state of knowledge as of about 2025, not on our objectives as of 2008. For the purpose of this planning exercise, between now and 2025 the following missions are assumed to have achieved their objectives: Mars Phoenix (en route to Mars as of this writing), MSL (scheduled for launch in 2011), an aeronomy orbiter (scheduled for launch in 2013), ExoMars (scheduled for launch in 2013), at least one more science orbiter, and the Mars Sample Return (MSR). Although other science missions will certainly be considered (most importantly, a network science mission), for the purpose of this planning we have not pre-judged NASA's decision-making and have only assumed the missions that seem most probable.

The results of the robotic missions between now and 2025 will answer some of the questions on our current horizon, which would therefore be removed and would be replaced by new questions; this is the scientific process. Although our ability to predict the results of these future missions and the kinds of new questions that will come up is partial, we do know the kinds of data that will be collected and the kinds of questions that these data are capable of answering. Thus, we can make some general projections of the state of knowledge as of 2025.

Goal I. DETERMINE WHETHER LIFE EVER AROSE ON MARS
By 2025, our assessments of habitability potential will be well advanced for some environments, particularly those that have been visited by the MSR or by major in-situ rovers with life-related experiments. However, it is likely that the habitability of the martian subsurface will be almost completely unexplored other than by geophysical methods. The objective relating to carbon cycling is likely to be partially complete, but in particular as related to subsurface environments. For the purpose of this planning, we assume that the investigations through 2025 have made one or more discoveries that are hypothesized as being related to ancient life (by analogy with the Allen Hills meteorite story, this is a particularly likely outcome of MSR). We should then be prepared for the following new objectives:

- Characterize the full suite of biosignatures for ancient life to confirm the past presence of life. Interpret its life processes and the origin of such life.
- Assess protected environmental niches that may serve as refugia for extant life forms that may have survived to the present. Find the life, measure its life processes.
- In earliest martian rocks, characterize the pre-biotic chemistry.

Goal II. UNDERSTANDING THE PROCESSES AND HISTORY OF CLIMATE ON MARS
By 2025, our objectives related to characterization of the Mars atmosphere and its present and ancient climate processes are likely to be partially complete. In addition to continuing long-term observations, our scientific questions seem likely to evolve in the following directions. Note in particular that if there is no robotic mission to one of the polar caps, the priority of that science is likely to be significantly more important than it is today because of the influence of polar ice on the climate system.

- Quantitative understanding of global atmospheric dynamics.
- Understand microclimates – range of variation, how and why they exist.
- Perform weather prediction.
- Understand the large-scale evolution of the polar caps including the modern energy balance, links with dust, carbon dioxide (CO_2), and H_2O cycles, changes in deposition and erosion patterns, flow, melting, age, and links between the two caps.

Goal III. DETERMINE THE EVOLUTION OF THE SURFACE AND INTERIOR OF MARS
As of 2006, there were two primary objectives within this goal: (1) Determine the nature and evolution of the geologic processes that have created and modified the martian crust and surface, and (2) characterize the structure, composition, dynamics, and evolution of the martian interior. These are broadly enough phrased that they are likely to still be valid in 2025. These two objectives, for example, currently apply to the study of the Earth, even after more than 200 years of geologic study by thousands of geologists. Given the anticipated robotic missions leading up to the first human missions, the first objective is likely to evolve in the following direction:

- Quantitatively characterize the different components of the martian geologic system (at different parts of martian geologic history), and understand how these components relate to each other.
- Understand the field context of the various martian features of geologic interest at both regional and local scale.
- Test specific hypotheses.
- Perform comparative planetology.

Unless a robotic geophysical network mission is scheduled before the first human mission, our progress on the second objective will be minimal and this will remain one of most important open questions.

3.1.3 Geology scientific objectives for the initial human exploration of Mars

Some of the most important questions about Goals I through III involve the relationship of H_2O to martian geologic and biologic processes as a function of geologic time. Mars has apparently evolved from a potentially "warm and wet" period in its early Noachian history to the later "cold and dry" period of the Amazonian period. Since rocks of different age are exposed in different places on Mars, understanding this geologic history requires an exploration program that also involves spatial diversity. One of the realities of geology-related exploration is that samples and outcrops are typically representative only of a certain geologic environment, and that acquiring information about other environments requires going to a different place. (A terrestrial analog would be asking how much we could learn about Precambrian granite by doing field work in the sedimentary rocks of the Great Plains.)

The absolute ages of surface units on Mars have been deciphered through indirect methods. Samples returned from the moon in the Apollo Program were used to provide constraints on the crater-size frequency distribution of the lunar surface (Gault, 1970; Hartman, 1972), and this has been applied to Mars, among other terrestrial planetary bodies (Barlow, 1988; Strom, 1992; Neukum, 2001). While this has provided a general history of martian surface processes, it does not allow for detailed study of specific martian periods, in particular the Hesperian and Amazonian periods when the impact flux greatly decreased. While martian meteorites have been analyzed and dated (Nyquist et al., 2001), not knowing their geologic context makes their incorporation into the geologic history of Mars difficult. While an MSR mission could potentially yield surface samples with known context, a robotic mission would not yield the array of optimal samples that would address a wide range of fundamental questions. A human mission might allow for greater access to samples that a robotic rover might not get to, and the capacity for real-time analysis and decision-making would ensure that the samples obtained that were would be the optimal available samples.

Human explorers would also have greater access to the near-subsurface of Mars, which would yield insights into climate and surface evolution, geophysics, and, potentially, life. Humans would be able to navigate more effectively through blocky ejecta deposits, which would provide samples that were excavated from great depth and provide a window into the deeper subsurface. Humans could trench in dozens of targeted locations and operate sophisticated drilling equipment that could drill to a depth of 500 to 1,000 meters below the surface. Our current understanding of the crust of Mars is limited to the top meter of the surface, so drilling experiments would yield unprecedented and immediate data. Drilling in areas of gully formation could also test the groundwater model by searching for a confined aquifer at depth.

We have analyzed three different exploration sites in detail as reference missions for the first program of human Mars exploration. The sites, which span the geologic history of Mars (one site for each period of martian history), allow for exploration traverses that would examine a variety of surface morphologies, textures, and mineralogy to address the fundamental questions posed by the MEPAG.

3.1.4 Geophysics scientific objectives for the initial human exploration of Mars

Mars geophysics science objectives fall into two broad categories: planetary scale geophysics (thousands of kilometers), and what might be called "exploration geophysics," which addresses regional (tens to hundreds of kilometers) or local scales (<10 km). The first category involves characterizing the structure, composition, dynamics, and evolution of the martian interior, while the second category addresses the structure, composition, and state of the crust, cryosphere, hydrologic systems, and upper mantle. Here we describe how these objectives might be met through investigations carried out on human missions. We assume here that no robotic missions to Mars before 2025 address the science issues in a complete way. For example, we assume that no network mission such as ML3N (National Research Council (NRC), 2006) will be flown. In general, Mars geophysics will be well served by the diversity of landing sites needed to pursue the geological and life-related objectives.

To characterize the structure and dynamics of Mars' interior band, we must determine the chemical and thermal evolution of the planet, including physical quantities such as density and temperature with depth, composition and phase changes within the mantle, the core/mantle boundary location, thermal conductivity profile and the 3-dimensional mass distribution of the planet. To determine the origin and history of the planet's magnetic field, we must discover the mineralogy responsible for today's observed remnant magnetization, and understand how and when the rocks bearing these minerals were emplaced. A key driver is the need to instrument the planet at appropriate scales: e.g., global seismic studies rely on widely separated stations so that seismic ray paths passing through the deep mantle and core can be observed. This need translates into multiple, widely separated landing sites for the first human missions. If only a single landing site is selected and revisited, far less information about Mars' interior will be obtained. A wide variety of exploration geophysics techniques could be brought to bear, including sounding for aquifers through electromagnetic techniques and reflection seismology to determine local structure. Magnetic surveys that are carried out at landing sites tell us about the spatial scales of crustal magnetization, and tie in to local and regional geology for context.

Geophysics measurement requirements span three disparate spatial scales, depending on the science that is to be done. At the largest scales (thousands of kilometers), characterizing the interior of Mars requires a widely spaced network of at least three emplaced central geophysics stations, one at each landing site. At regional scales (tens to thousands of kilometers), characterizing crustal structure, magnetism, and other objectives requires mobility to emplace local networks around a landing site. Finally, at local scales (10 km), mobility is key to performing traverse geophysics, and in carrying out investigations (such as seismic or electromagnetic sounding) at specific stations along a traverse. The central geophysics stations and the regional scale networks would be emplaced and left to operate autonomously after the human crew departs. Traverse and station geophysics would be carried out only during the human mission, unless this could be done robotically after completion of the human mission.

Central geophysical stations at each landing site would include passive broadband seismic, heat flow, precision geodesy, and passive low-frequency electromagnetic instrumentation. Satellite geophysics stations would include the nodes of a regional seismic array and vector magnetometers. Along the traverses, experiments would be performed at sites of interest. These would include active electromagnetic (EM) sounding for subsurface aquifers, active seismic profiling to establish structure with depth, and gravity measurements. Ground-penetrating radar and neutron spectroscopy along the traverse track help map out subsurface structure and hydration state/ice content for the near-subsurface.

3.1.5 Atmosphere/climate scientific objectives for the initial human exploration of Mars

In the human era of exploration, atmospheric measurements at all sites would be seen as important not only to understanding Mars' atmosphere and climate and to planning human surface operations, but also as an environmental characterization that is essential to the interpretation of many life and geology objectives. The trend towards system science called out in MEPAG (2006) as a "ground-to-exosphere approach to monitoring the martian atmospheric structure and dynamics" will continue with more emphasis on the mass, heat, and momentum fluxes between the three Mars climate components: atmosphere, cryosphere, and planetary surface.

This systems approach will be enabled by advances in Mars global circulation models (GCMs), a doubling in length of the global time-series that is derived from monitoring Mars' surface and atmosphere from orbit, new atmospheric vertical structure information from the Mars Express and MRO, new anticipated global data sets on aeronomy, and atmospheric composition and winds; and by network science and coordinated lander-orbiter campaigns, such as that planned with Phoenix-MRO. In 2007, trends in Mars GCM development are towards a coupling of the upper and lower atmosphere, coupling with regolith models, and integrating models of atmospheric chemistry and dynamics, multiscale, nested models – where small-scale surface-atmosphere interactions can be studied within the context of global transport and data assimilation. Models have not yet been successful in reproducing the observed martian dust cycle with active dust transport. Temperature and wind profile information from heights between the top of instrumented masts and the free atmosphere will likely remain sparse or nonexistent.

Understanding Mars' past climate will benefit from anticipated new knowledge of current atmospheric escape rates that will be gained from the 2013 Mars Aeronomy Scout. However, a significant advancement in the key area of access to the polar stratigraphic record is not expected in the decades before human exploration. In 2030, this will therefore remain one of the highest priorities for MEPAG. On the other hand, the study of the paleoclimatic parameters that are imprinted in the ancient geological record (e.g., Noachian to Amazonian periods) also concerns the high priorities of the MEPAG, which directly relates to unlocking the ancient climatic conditions of Mars through a physical (e.g., geomorphic and/or sedimentary), petrological, mineral, and geochemical (including isotopic) material characterization.

The emphasis of atmospheric science measurements by human missions would likely focus on processes within the planetary boundary layer (PBL), which is surface to 2 km, where surface-atmosphere interactions impart fundamental influences on the dynamical, chemical, and aerosol characters of the global Mars atmosphere. For the PBL, all spatial scales are important in turbulent exchange, from centimeters to kilometers, in both horizontal and vertical dimensions. It is the wide diversity of spatial scales and the driving importance of the near-surface contribution that lead to fundamental limitations of orbital remote-sensing: surface field campaigns are still a major thrust of atmospheric boundary layer research on Earth for understanding small-scale variability. Through nonlinear processes, small-scale variability can significantly influence the global climate. Human atmospheric observations could provide optimum in-situ and remote access to the PBL and, in turn, characterize local environmental conditions in support of human operations.

Atmospheric dynamics. This is important because it determines the basic thermal structure of the martian atmosphere, the global transport of volatiles (CO_2, H_2O, dust), and the maintenance of the martian polar ice caps, all of which vary on seasonal and inter-annual timescales. Although significant progress in our understanding would have been made by the time of the first human missions, observational constraints are likely to remain sparse, particularly for the near-surface atmosphere. This reflects both the limitations of orbital remote sensing and the limited number of expected robotic lander/rovers. Dedicated observations of surface pressure and temperature-wind-dust profiles of the PBL from distributed surface stations constitute a key priority for human investigations of Mars' atmospheric dynamics.

Atmospheric Dust. Atmospheric heating that is associated with atmospheric dust intensifies global atmospheric circulation and near-surface winds, which in turn increases lifting of surface dust into the atmosphere. A dramatic result of this dust radiative-dynamic feedback is ubiquitous aeolian activity on Mars, with significant dust lofting and transport occurring over a wide range of spatial and temporal scales. These range from nearly continuous dust devil activity, to regional dust storms in every Mars year, to global dust storms that may occur once every three or four Mars years (Cantor et al., 2001). As a consequence, atmospheric dust plays a major role in the spatial and seasonal Mars atmospheric thermal structure and circulation as well as in its variability. It remains uncertain whether global surface dust distributions limit or are influenced by atmospheric dust transport. In-situ observations of dust surface flux (lifting and deposition), particle sizes, radiative properties, and vertical profiles within the PBL constitute primary objectives for human atmospheric dust studies. In addition, understanding the dust-induced atmospheric density scale height variability is an important issue for proper development of future entry, descent, and landing systems.

Atmospheric Water: Atmospheric H_2O, in the form of vapor and ice clouds, plays significant roles in atmospheric chemistry, dust radiative forcing, and climate balance. The photolysis products of atmospheric H_2O vapor determine Mars trace species abundances (Nair et al., 1994). Water ice clouds have long been associated with major topographic features, autumnal polar hoods, and a variety of cloud wave structures (Kahn, 1984). The existence of an aphelion, low-latitude cloud belt is identified as a significant influence on the vertical distribution of atmospheric dust and H_2O vapor (Jakosky and Farmer, 1983), as well as meridional transport of atmospheric H_2O (Clancy et al., 1996). Atmospheric exchange with polar cap H_2O ice deposits dominates the seasonal variation of atmospheric H_2O vapor, whereas atmospheric exchange with subsurface ice and adsorbed H_2O at lower latitudes remains uncertain. Human investigations of atmospheric H_2O are likely to focus on vertical profile measurements within the PBL, which are not easily addressed from orbital remote sensing. Subsurface core sampling of adsorbed H_2O and H_2O-ice H_2O deposits (site-dependent in this case) also constitutes a key Mars H_2O objective that is uniquely facilitated by human measurements.

Atmospheric Chemistry: The trace chemical composition of the current martian atmosphere reflects the photochemical cycles that are associated with the major atmospheric constituents CO_2, H_2O, and nitrogen (N_2); and perhaps nonequilibrium chemistry that is associated with potential subsurface sources – sinks of methane (CH_4), sulfur dioxide (SO_2) and hydrogen peroxide (H_2O_2) (Levine, 1985; Yung and DeMore, 1999). Some of these compounds can be essential to sustain a Mars cryptic biosphere through direct or indirect (biochemical pathways (e.g., atmospheric oxidants can be used as electron acceptors for microbial metabolism, whereas reducing gases such as –CH_4- can be electron donors). Definitions of spatial and seasonal variations in atmospheric trace composition remain tentative. The most problematic trace species measurements, on both observational and modeling grounds, are the recent reported detections of significant atmospheric methane abundances (Formasano et al., 2004, Krasnopolsky et al., 2004). Human observations of atmospheric chemistry are likely to focus on detections of locally enhanced CH_4, SO_2, hydrogen sulfide (H_2S), hydrogen cyanide (HCN), or H_2O_2 concentrations that are associated with confined source regions that are specific to the geology, geophysics, or life site.

Electrical Effects: Experimental and theoretical investigations of frictional charging mechanisms in both small- and large-scale meteorological phenomena suggest that Mars very likely possesses an electrically active atmosphere as a result of dust-lifting processes of all scales, including dust devils and dust storms. Electrical effects impact human exploration and the environment of Mars as a source of both continual and episodic energy. Differential charging in the presence of electrified dust between separate objects that then come into contact and cause a discharge would directly damage electronics or interfere with radio communications. Dust adhesion may also be dominated by electrical effects, with implications in terms of its transport into the habitat/human environment where other effects may take over (toxicity, friction in seals/machinery, etc.). Currently, measurements of electric charging within the Mars atmosphere do not exist. For operational safety concerns alone, basic measurements of martian surface charging conditions should be obtained prior to human exploration activities. Human measurements of atmospheric charging within active dust devils are especially relevant to the dynamic response times associated with dust devil occurrences and motions.

3.1.6 Biology/life scientific objectives for the initial human exploration of Mars

Human-enabled biological investigations on Mars would focus on taking samples and making measurements to determine whether life ever arose on Mars. This goal is consistent with the 2006 MEPAG goals and priorities, and we do not see this goal changing within the next 30 years.

The search for life on Mars can be generally broken into two broad categories: (1) the search for evidence of past life on Mars, which may or may not still be alive; and (2) the search for present (extant) life. Both have been, and will continue to be, based on a search for H_2O, since all life on Earth requires H_2O for survival. Abundant evidence on the martian surface of past H_2O activity (e.g., rivers, lakes, groundwater discharge) has led to Mars becoming a strong candidate as a second planet in our solar system with a history of life. With our increasing knowledge of the extremes under which organisms can survive on Earth, especially in the deep subsurface, whether martian life is still present today has become a compelling and legitimate scientific question.

The NRC was recently commissioned to do a study to develop "an up-to-date integrated astrobiology strategy for Mars exploration that brings together all the threads of this diverse topic into a single source for science mission planning." This NRC report, which was published in 2007, is entitled "An Astrobiology Strategy for the Exploration of Mars (NRC, 2007). Although this report did not consider how to do science with humans, we rely heavily on it and earlier MEPAG documents here to provide snapshots of the current community thinking on astrobiological investigations on Mars.

As pointed out by the NRC (2007), the search for life on Mars requires a very broad understanding of Mars as an integrated planetary system. Such an integrated understanding requires investigation of the following:

1. The geological and geophysical evolution of Mars;
2. The history of Mars' volatiles and climate;
3. The nature of the surface and the subsurface martian environments;
4. The temporal and geographical distribution of H_2O;
5. The availability of other resources (e.g., energy) that are necessary to support life; and
6. An understanding of the processes that control each of the factors listed above.

Many of these investigations are well under way robotically and will be much further advanced through future robotic missions and sample return missions.

3.1.7 The search for extant life

The NRC (2007) suggests a number of high-priority targets based on evidence for present-day or geologically recent H_2O near the surface. These targets are

1. The surface, interior, and margins of the polar caps;
2. Cold, warm, or hot springs or underground hydrothermal systems; and

3. Source or outflow regions that are associated with near-surface aquifers that might be responsible for the "gullies" that have been observed on the martian surface.

The MEPAG Special Regions Science Analysis Group (2006) noted that the sites where recent H_2O may have occurred might also include some mid-latitude deposits that are indicative of shallow ground ice. Conditions in the top 5 m of the martian surface are considered extremely limiting for life. Limiting conditions include high levels of ultraviolet radiation and purported oxidants as well as most of the surface being below the limits of H_2O activity and temperature for life on Earth. For these reasons, finding evidence of extant life near the martian surface will likely be difficult, and the search will almost certainly require subsurface access. This was also a key recommendation of the NRC (2007).

3.1.8 The search for past life

The NRC (2007) lists sites that are pertinent to geologically ancient H_2O (and, by association, the possibility of past life), including the following:

1. Source or outflow regions for the catastrophic flood channels;
2. Ancient highlands that formed at a time when surface H_2O might have been widespread (e.g., in the Noachian); and
3. Deposits of minerals that are associated with surface or subsurface H_2O or with ancient hydrothermal systems or cold, warm, or hot springs.

3.2 Objectives Related to Preparation for Sustained Human Presence (Goal IV+)

MEPAG's Goal IV is interpreted to be related to preparation for the first human explorers, so by definition, it will be complete before the initial set of human missions has been attempted or the activity will have been shown not to be necessary. We refer to Goal IV+ as the preparation for the sustained human presence on Mars beyond that of the DRA 5.0 mission set. Specific objectives within Goal IV+ could be carried out either within the context of the DRA 5.0 missions, or by the preceding robotic program. The scope of the representative scenarios includes developing the knowledge, capabilities, and infrastructure that are required to live and work on Mars, with a focus on developing sustainable human presence on Mars.

Major Goal IV+ objectives decomposition

The four major Goal IV+ objective areas are: Mars Human Habitability/In-Situ Resource Utilization (ISRU), Exploration Systems Development, Operational Capabilities, and Other. Within each of these areas are multiple categories of lower-level objectives, as shown in table 3-2.

Table 3-2. Goal IV+ Objective Decomposition

Mars Human Habitability ISRU	Exploration Systems Development	Operational Capabilities	Other
Human Health	General Infrastructure	Crew Activity Support	Planetary Protection
Environmental Characterization	Operational Environmental Monitoring		Historic Preservation
Environmental Hazard Mitigation	Life Support		Commercial Activities
Mars Resource Utilization	Habitation Systems/ISRU		Global Partnership
	EVA Systems		Public Engagement
	Power		
	Communications		
	Position, Navigation, and Time		
	Transportation		
	Surface Mobility		
	Operations, Testing, and Verification		

3.2.1 Sustainability-related objectives for the initial human missions to Mars

Within each of the four major Goal IV+ objective areas, the Goal IV+ study team defined the projected Mars exploration objectives for human missions one through three. These are as follows:

1. *Habitability*: There are three resulting objectives in the Mars Human Habitability/ISRU area. The first objective is to develop the capability of providing crew needs from local resources. An example of this is in-situ food production. The second objective is to develop the capability of extracting power and propulsion consumables from local resources. This could be accomplished through ISRU processing of the martian atmosphere or regolith to produce methane or of other chemicals needed for power and propulsion technologies. The third objective is to develop and test the capabilities needed for in-situ fabrication and repair. This could be accomplished by fabricating infrastructure element replacement parts on the martian surface from raw materials brought from Earth, or by reusing parts from other infrastructure elements that are no longer in use (e.g., a decent stage that is used only for landing on the surface).

2. *Systems Development*: The Exploration Systems Development area includes three objectives, all of which relate to the establishment of reliable and robust space systems that would enable gradual and safe growth of capabilities. The first such capability is the number of individuals that can be supported by the infrastructure on Mars. The exploration systems that are developed would also work to increase the duration of time during which individuals can live safely on the planet. Another thrust would be the gradual increase in the range of mobility that is provided to visiting crews. As each of these capabilities is realized and matured, the potential for even greater exploration, science, discovery, and new technology is greatly enhanced.

3. *Self-sufficiency*: The level of self-sufficiency of operations for Mars missions also must increase and, hence, is the objective in the Operational Capabilities area. Due to the complexity of procedures and the communications delay, among other factors, a crew that is operating on the surface of Mars would need to be independent from the supporting personnel who are located back on Earth. These Earth-based teams would of course be available to offer assistance in nonemergency situations. However, the new complications of a martian mission warrant consideration of a day-to-day level of autonomy that is not currently present in space shuttle and ISS missions.

4. *Other Objectives*: The study resulted in three "Other" objectives, which address planetary protection concerns, partnerships, and public engagement. Insofar as these are concerned,
   - Special care must be taken not to contaminate the natural environment, where scientific measurements would require pristine samples, as well as any areas to which the human crews would be exposed so as to protect their health.
   - A sustained human presence on Mars would require the development of partnerships. Promoting agreements and collaboration among governmental, international, commercial, and other entities would be a necessary challenge.
   - Another objective in this area is to provide and sustain public engagement. The exploration of and sustained human presence on Mars would be obviously be a grand undertaking, one that requires long-term, continual public support.

3.3 *Objectives Related to Other Classes of Science (Goal V)*

Potential science objectives that are appropriate to the initial human missions to Mars extend beyond those relating solely to the scientific exploration of Mars as a planet or the preparation for a sustained human presence on Mars. As a unique planetary specimen, Mars is relevant to the study of the entire solar system, including its evolution under the influence of the sun (Heliophysics), and to the study of the solar system as an important specimen of stellar evolution (Astrophysics), as well as other science disciplines. In addition, Mars may be a unique location from which to perform certain astrophysical observations.

3.3.1 Heliophysics of Mars' environment

The martian system, as an archive of solar system evolution (space climate) and a case of planetary interfaces responding to immediate solar influences (space weather), is of great interest to the science of Heliophysics. These influences range from solar irradiance and high-energy particles irradiating the planet's surface, to solar wind and magnetic fields driving disturbances of the martian atmosphere and ionosphere. Mars also represents an important key instance of fundamental Heliophysical processes that influence the habitability of planets. Because the space environment matters to the safety and productivity of humans and their technological systems both at Mars and in transit, it is essential that we monitor Heliophysical conditions between Earth and Mars and understand solar effects on the martian atmosphere, which are relevant for vehicles in Mars orbit or traveling through the atmosphere to the surface environment. An important supporting objective is to understand the influence of planetary plasmas and magnetic fields and their interaction with the solar wind plasma.

3.3.2 Space weather

The sun and interplanetary medium permeating our solar system, as well as the universe at large, consist primarily of plasmas. This leads to a rich set of interacting physical processes and regimes, including intricate exchanges with the neutral gas environments of planets. In preparation for travel through this environment, human explorers must anticipate and prepare for encounters with hazardous conditions stemming from ionizing radiation. We must develop mitigation strategies and a complete understanding of the many processes that occur with such a wide range of parameters and boundary conditions within these systems. We must be able to predict the behavior of the complex systems that influence the hazardous conditions that crews would encounter. Hazards in planetary environments must be understood, characterized, and mitigated. We must also understand how space weather impacts the planetary environments that affect exploration activities, from spacecraft staging in low Earth orbit, to transfer orbits, on through entry, descent, and landing (EDL) at Earth and Mars. Reliable communications and navigation for spacecraft and surface crews would require improved understanding of the ionospheres of both Earth and Mars. Although the sun and its variability drive these environments, internal response processes must also be understood. Among the many questions to be answered, the following are perhaps the most significant: What are the mean conditions, variability, and extremes of the radiation and space environment for exploration of Mars? How does the radiation environment vary in space and time, and how should it be monitored and predicted for situational awareness during exploration? What is the relative contribution from solar energetic particles and cosmic radiation behind the various shielding materials that are used and encountered, and how does this vary?

3.3.3 Laser ranging for astrophysics

While observations from free space offer the most promise for significant progress in broad areas of astrophysics, some investigations could be uniquely enabled by the infrastructure and capabilities of a human mission to Mars. Among the most promising in this respect are laser ranging experiments to test a certain class of alternative theories (to general relativity) of gravity. Such experiments become even more valuable when considered in the context of a humans-to-Mars architecture. The long baseline measurements that are afforded by laser ranging from Mars provides a unique capability that would otherwise not be enabled by free space implementations or via a lunar architecture.

3.4 Goals and Objectives Summary Implications

During the development of the Mars Design Reference Architecture 5.0, the HEM-SAG was given options to consider that were based on the accumulated body of previous human Mars mission studies. These options were developed to provide a better understanding of the relationship between the various exploration goals and objectives and resulting implementation approaches of meeting those goals. Deliberations by the HEM-SAG resulted in the following summary implications:

- *For the first three missions, three different sites or the same site?* Over the last decade, exploration of Mars by robotic orbiters, landers, and rovers has shown Mars to be a planet of great diversity and complexity. This diversity and complexity offers a unique opportunity for humans on the surface of Mars to obtain data and measurements that could not be obtained by robotic probes alone. To use human explorers effectively in addressing these scientific questions, the first three human missions to Mars should be to three different geographic sites. The Goal IV+ objectives lend themselves best to repeated visits to a specific site on Mars, however. Repeated site visits would enable a buildup of infrastructure that would benefit the longer-term missions of the Goal IV+ objectives. This buildup would provide more systems for use by the crews such as habitable volume, mobility aids, and science equipment. These systems and the potential for spares could also potentially reduce the amount of logistics required for the long-term missions.
- *Short stay (30 days) or long stay (500 days)?* It is clear that productivity of the missions is amplified many-fold in a 500-day scenario as compared to a 30-day scenario. This is particularly true of scientific objectives that are related to geology and the search for life, for which we need to maximize the amount of time that the astronauts spend examining the rocks and the diversity of the samples that are collected. Longer stays allow for a more comprehensive characterization of certain environmental parameters and a longer baseline of measurements. This specific and long-duration knowledge will be essential in the development of health monitoring and hazard mitigation strategies for both the crew and infrastructure elements. The systems required for long stays are also more supportive of the eventual longer term missions that would achieve sustained human presence
- *Mobility.* Achieving these scientific objectives would require mobility. Although different possible landing sites have different spatial relationships, it is possible to estimate that the capability of traveling a radial distance of several hundred kilometers would allow a full range of landing site options.
- *Subsurface access.* It is possible that drilling depths in the range of 100 to 1000 m would be necessary, depending on the drilling site and the goal of the drilling.
- *Returned sample science.* Since human missions to Mars have a round-trip component to them, they naturally lend themselves to returned sample science. To maximize the value of the returned sample collection, it would be necessary to have a habitat laboratory for two purposes: (1) to help guide the on-Mars field strategies, and (2) to ensure the high grade of the samples to be returned. Sample conditioning and preservation will be essential. The minimum mass of samples to be returned to Earth is to be determined, but it could be as much as 250 kg.
- *Instruments that operate after humans leave.* Several types of monitoring stations should be configured so that they can continue operating after the astronauts leave. This would specifically include network stations for seismic monitoring and long-duration climate monitoring.
- *Planetary protection.* The impact of human explorers and potential "human contamination" of the martian environment in the search for present-day life on Mars is a problem that requires more study and evaluation, and that must be solved prior to the first human landing on Mars.

Given that the engineering of missions to Mars are constrained to be either "short stay" or "long stay" (see Section 6), and assuming that the initial human exploration of Mars consists of a program of three missions, a key tradeoff is mission duration and whether the missions are sent to the same or to different sites. **From the perspective of our scientific goals, it is clear that our progress would be optimized by visiting multiple sites and by maximizing the stay time at those sites.** The same argument regarding diversity of sites was raised, and followed, during the Apollo Program. The longer stay time is needed because the geology of Mars at many sites has complexities that would take a significant amount of time to resolve. If we are to bring the unique attributes of human explorers to bear, we would need to give them enough time on the outcrops.

4 TRANSPORTATION SYSTEMS

The technical assessments conducted for the DRA 5.0 focused primarily on the launch vehicle, interplanetary transportation, and EDL systems. Assessments of the applicability of the Orion crew exploration vehicle (CEV) as well as the Mars DAV and the interplanetary transit habitat were also conducted, but not to the same level of detail. Assessments of using the Constellation Program's (CxP's) heavy-lift launch vehicle (HLLV), the Ares V launch vehicle, for a human mission to Mars were examined both in the context of the required performance (e.g., initial mass in low-Earth orbit (IMLEO), number of launches, etc.) and in the context of their impacts to existing ground infrastructure at KSC. For the in-space transportation system for crew and cargo, the design team assessed nuclear thermal and advanced chemical propulsion,

Advanced Propulsion – Depiction of NTR propulsion Mars transfer vehicle in LEO prior to departure. Glenn Research Center 2007.

and determined that the NTR was the preferred approach, while retaining chemical/aerocapture as a backup option. In previous design reference missions (DRMs), a small capsule was envisioned for the Earth return vehicle (ERV), but with the design of the Orion CEV there is now a block-upgrade path that would seek to augment the capsule that is currently being designed to go to the moon for use on a round-trip Mars mission. This would primarily involve upgrading the Thermal Protection System (TPS) on the current Orion design to account for the higher Earth entry speeds and certifying the vehicle for extended dormant times in a space environment. Perhaps the most important advancement in knowledge since the last reference mission comes with respect to the EDL systems that are to be employed at Mars to land payloads on the order of 30 to 50 t. Previous estimates of human-class EDL system mass were determined to be optimistic given the great unknowns that are still associated with landing robotic payloads greater than 1 t on Mars. Additional knowledge and insights that were gained with the successful robotic mission EDL designs of the last decade (Mars Pathfinder, the MERs *Spirit* and *Opportunity*, and the Mars Phoenix lander) have also resulted in more realistic estimates for EDL system masses required for robust EDL system designs. The new assessment details a more conservative estimate of EDL system mass, which has substantially increased, even in spite of the advantage gained from the presumed use of a common Ares V launch shroud/aeroshell payload entry shield. Mass increases in this subsystem are a prime contributor to the overall increase in the initial mass to LEO estimates given in this DRA as compared to previous DRMs. Despite the fact that detailed analysis work was not performed during DRA 5.0 on the MTV, Mars ascent vehicle, or SHAB lander, past analysis of all three vehicles was updated with current assumptions. This especially applies to the case of the Mars ascent vehicle in which ascent stages using ISRU were parametrically sized in comparison to ascent stages that were fully fueled from the beginning. The impact of using ISRU on the Mars ascent vehicle was traced back all the way to LEO to make a recommendation with regards to the use of ISRU.

4.1 *Interplanetary Trajectory and Mission Analysis*

Although no date has been chosen for the first human mission to Mars, for this study high-thrust trajectories were analyzed for round-trip crewed missions to Mars with Earth departure dates ranging from 2030 to 2046. These dates were chosen to assess the variability of mission opportunities across the synodic cycle, and not to represent proposed actual mission dates. Mission opportunities occur approximately every 2.1 years in a cycle that repeats every 15 years (the synodic cycle). (The trajectories from one 15-year cycle to the next do not match exactly, but are very similar and sufficient for initial planning purposes. The duration required for a more exact match is 79 years.) Along with the crewed missions, one-way cargo delivery trajectories were also generated that depart during the opportunity preceding each crewed mission. Each cargo mission delivers two vehicles to Mars.

The trajectories that would be used for human crews balance low interplanetary trip times with the cost (i.e., propellant) of achieving the missions. This is facilitated by allowing long Mars stay times. For each opportunity, the outbound and inbound transit times are minimized such that desired departure energies (determined by V_∞) are not exceeded. Again, the Mars stay time is allowed to vary so that the lowest interplanetary flight times are possible. The supporting cargo flights follow minimum energy trajectories with no restriction on the outbound transit times. The cargo departures occur approximately 2.1 years before each crew mission. This allows confirmation that the cargo elements have successfully reached their destinations and are functioning properly before the crew leaves Earth.

In this analysis, all vehicles depart from a 407-km circular orbit, and a two-burn Earth escape is performed to reduce the gravity loss penalties. At Mars, the vehicles are inserted into a 1-sol orbit (250 km × 33,793 km). Both propulsive and aerocapture cases were investigated for the cargo missions, while for the crewed vehicles only propulsive orbital insertions were considered. Further discussion on the architecture and mission trades that were conducted is provided in Section 6 of this report.

Representative trajectories for the cargo and crew missions for an example 2037 crew mission are shown in figure 4-1. The displayed crewed profile corresponds to the all-propulsive 2037 opportunity with transit times of 174 days outbound and 201 days inbound. The Mars stay time is 539 days, and the total mission duration is 914 days. Again, note that the majority of the mission duration is spent on the surface of Mars, while the interplanetary transit times are reduced to minimize the exposure of the crew to harmful solar and galactic cosmic radiation. The supporting cargo vehicle departs Earth a little more than 2 years before the crewed mission in 2035 and follows a minimum energy trajectory. The trip time of 202 days is the quickest cargo flight time that was observed over the dates analyzed.

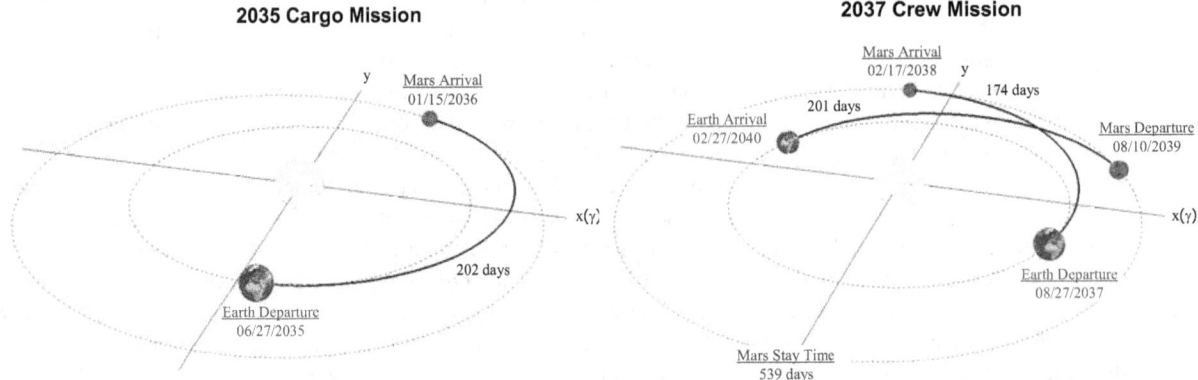

Figure 4-1. Cargo and crew trajectories for the example 2037 mission.

The crew and cargo mission delta-Vs over the dates of interest are shown in figure 4-2. For the crewed missions, all TMI maneuvers are designed to achieve the maximum allowed Earth departure V_∞, except for the 2031 and 2035 opportunities. As the time of flight is reduced in the 2031 case, the maximum allowed Mars arrival V_∞ is exceeded before the Earth departure limit is reached. The 2035 case includes a 180-day outbound trajectory. In fact, the flight time could be reduced to around 140 days before violating the end-point constraints. The longer flight time is shown to indicate that, if desired, more crewed payload could be delivered with a reasonably fast transit. Fewer restrictions are placed on the trajectories for the cargo missions. The flight times are allowed to vary to minimize the total effect of the TMI and MOI (in the all-propulsive missions) maneuvers. The all-propulsive vehicle design is determined by the worst-case delta-Vs, which for the TMI maneuver occurs in 2037 while the 2030 opportunity contains the maximum MOI requirement. In general, the aerocapture TMI variation resembles that of the propulsive MOI cases with slightly lower values. This is due to the relaxed effect of the Mars arrival velocity. The 2043 cases appear to contradict this statement, but the aerocapture case represents a fast-transit time (240 days) that is feasible because no propellant is required for MOI. If the aerocapture vehicle follows a longer trajectory, the TMI delta-V is less than or equal to that of the all-propulsive mission.

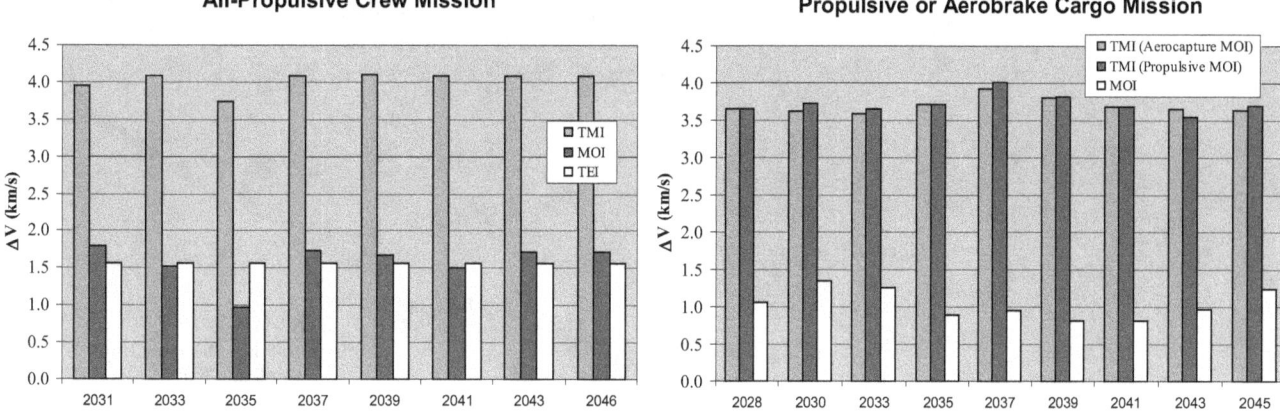

Figure 4-2. Trajectory delta-V summary for the reference conjunction class missions.

4.2 Heavy-Lift Launch Vehicle

The reference HLLV that is currently envisioned for NASA's human lunar return (HLR) is called Ares V by the CxP. The Ares V design continues to evolve, and the 45 series configuration (figure 4-3) served as the point of departure for the Mars DRA 5.0 study. It consists of two 5-segment reusable solid rocket boosters (RSRBs), a core stage that is powered by five Pratt & Whitney Rocketdyne RS-68B engines, an Earth departure stage (EDS) powered by one Pratt & Whitney Rocketdyne J-2X, and a payload shroud. This vehicle has a gross liftoff mass of approximately 3,323 t (7,326 klb$_m$) and a height of 110.3 m (361.9 ft). Because a new follow-on HLLV that was specifically designed for Mars would be too expensive, emphasis was placed on analyzing how well the various Ares V design options that are currently being designed for the lunar mission could be adapted to meet the mission objectives for Mars. As the Ares V design evolves for the human lunar return mission, its capabilities and performance must be continually assessed as to its suitability to meet key Mars mission requirements.

During the Mars architecture study, several different shroud configurations were examined to determine the effect of the shroud dimensions and delivery orbit on overall architecture performance on not only the launch vehicle but the shroud influence on the interplanetary transportation system, the EDL system, as well as other mission payloads. The shroud dimensions investigated ranged from 8.4 to 12 m in diameter and 12 to 35 m in length. In addition, the concept of a dual-purpose shroud that would be used for both the launch to LEO and Mars atmospheric entry (i.e., reinforced with TPS for EDL) was examined. The length of this dual-use shroud was defined as 30 m, including the transition cone with an outer diameter of 10 m.

The performance curves for the three different shroud sizes that were considered in this study are provided in figure 4-4. Specific launch vehicle performance for both the standard (Option B) and dual-use shroud options to the reference 407-km circular reference orbit that was chosen by the architecture team is also indicated. These curves provide the gross performance to the desired orbit. The net performance that is available for the actual payloads must account for items such as payload adapters, performance margins, and airborne support equipment. These accommodations would reduce the payload that is available by approximately 8 to 10% of the gross performance. In addition, the dual-use shroud option must accommodate the structural aeroshell, which was estimated to be approximately 40 t.

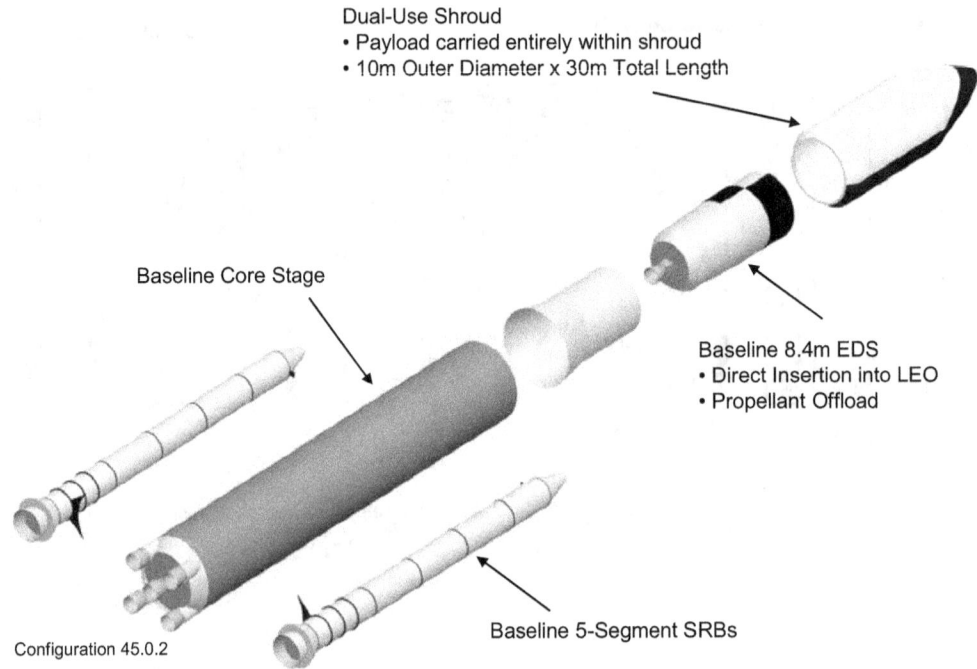

Figure 4-3. Reference Ares V launch vehicle.

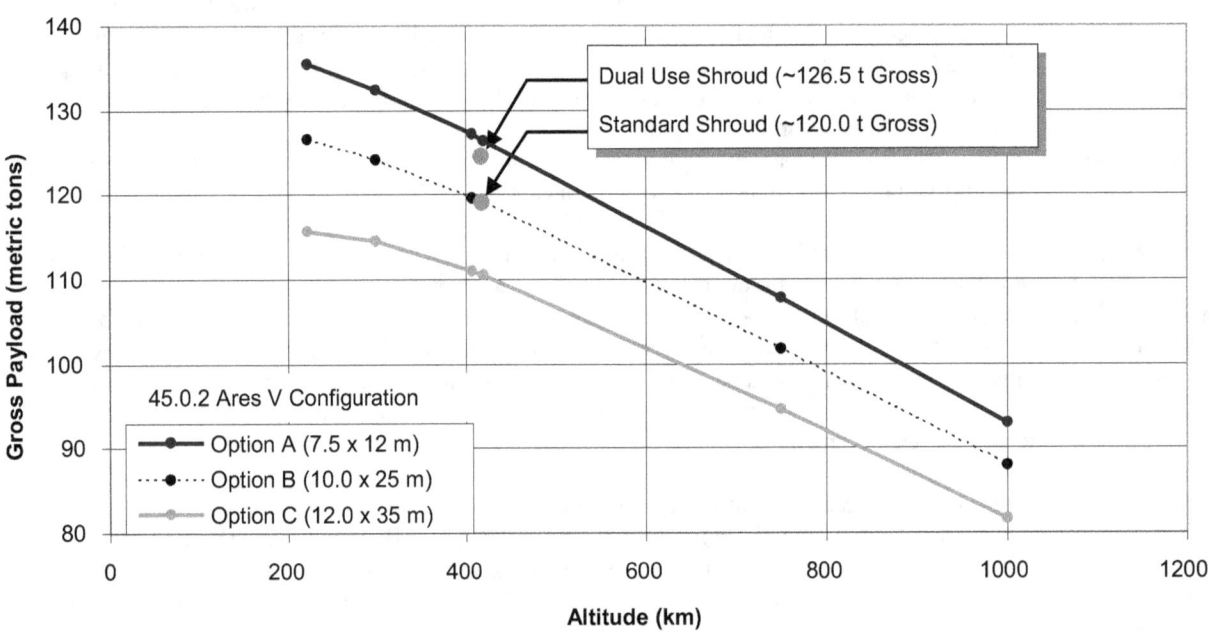

Figure 4-4. Launch vehicle payload performance (45.0.2 configuration).

After completion of this Mars Architecture study, a new series of launch vehicle configurations was assessed to improve the performance for the lunar missions (these were assessed during the CxAT_Lunar study and are referred to as the 51 series Ares V launch configurations). Although not addressed in detail by the MAWG, it is anticipated that further improvements in Ares V performance and payload size could be realized that would benefit the Mars architecture to help reduce the number of launches. Future Mars architecture studies and detailed assessments should help refine mission and system requirements and help ensure an Ares V that could support both NASA's lunar and Mars mission scenarios. In addition, further assessments regarding manufacturing, assembly, integration, test, and checkout of the systems that are required for the higher launch rate that is associated with the Mars missions, as compared to the lunar missions, is warranted.

4.3 In-Space Transportation: Nuclear Thermal Rocket Reference

During development of DRA 5.0, the design team conducted top-level performance assessments of both the NTR and advanced chemical propulsion. Based on the assessments that were conducted, the team concluded that the NTR was the preferred transportation technology for both the crew and the cargo vehicles and, thus, should be retained as the reference vehicle, with chemical/aerocapture as an option. The NTR is a leading propulsion system option for human Mars missions because of its high thrust (10's of klbf) and high specific impulse (Isp 875–950 s) capability, which is twice that of today's liquid oxygen (LOX)/liquid hydrogen (LH_2) chemical rocket engines. Demonstrated in 20 rocket/reactor ground tests during the Rover/Nuclear Engine for Rocket Vehicle Applications (NERVA) Programs, the NTR uses fission-reactor-generated thermal power rather than chemical combustion of an oxidizer-fuel mixture to directly heat LH_2 propellant for rocket thrust. NASA's previous Mars DRM studies, DRM 3.0 in 1998 and DRM 4.0 in 1999, used a "common" propulsion module with three 15,000 klbf NTR engines. The use of clustered, lower-thrust (15–25 klbf) engines provides an "engine-out" capability that could increase crew safety and reduce mission risk. The time and cost to develop and ground test these smaller engines is also expected to be less then that required for higher-thrust engines. Both conventional NTR engines (thrust only) and bimodal nuclear thermal rocket (BNTR) engines, which are capable of producing both thrust and modest amounts of electrical power (few 10's of kWe) during the mission coast phase, were examined in addition to zero-gravity and artificial gravity (AG) crewed MTV design concepts. The current Mars DRA 5.0 study efforts considered "thrust-only" NTR engines, zero-gravity crewed MTV designs, and photovoltaic arrays (PVAs) to supply spacecraft electrical power.

The cargo and crewed NTR MTV concepts that were developed for the long surface stay "split mission" DRA 5.0 are shown in figure 4-5. All vehicles use a common "core" propulsion stage with three 25-klbf NTR engines to perform all of the primary mission maneuvers. In-line and jettisonable drop tanks augment the core stage LH_2 propellant load for the different vehicles as needed. The propulsion stage carries circular Orion-type PVAs for auxiliary electrical power to run key stage subsystems (e.g., zero boil-off (ZBO) LH_2 cryocoolers) as well as a storable propellant Reaction Control System (RCS) for Earth orbit AR&D of MTV components and for orbit maintenance during the LEO loiter phase.

Two cargo flights are used to pre-deploy a cargo lander to the surface and a habitat lander into Mars orbit where it remains until the arrival of the crewed MTV during the next mission opportunity. Five Ares-V flights, which are carried out over 120 days, are required for the two cargo vehicles. The first two Ares-V launches deliver the NTR core propulsion stages while the third launch delivers the two short "in-line" LH_2 tanks that are packaged end-to-end. Once in orbit, the in-line tanks separate and dock with the propulsion stages, which function as the active element during the AR&D maneuver. The two aerocaptured payload elements are delivered on the last two Ares-V launches.

Each cargo vehicle has an IMLEO of 246.2 t and an overall length of 72.6 m, which includes the 30-m-long aerocaptured payload. The total payload mass (aeroshell, EDL system, lander descent stage, and surface payload) is 103 t, which is consistent with a surface strategy using nuclear power and ISRU. The NTR propulsion stage has an overall length of 28.8 m (26.6 m with retracted nozzles for launch) and a launch mass of 96.6 t. The stage LH_2 tank has an inner diameter of 8.9 m and a propellant capacity of 59.4 t. The short in-line tank has a launch mass of 46.6 t and an overall length of 13.3 m including the forward and rear adaptor sections, and it holds 34.1 t of LH_2. Each NTR cargo vehicle also carries 5.2 t of RCS propellant, which is used for LEO operations, coast attitude control, mid-course correction, and Mars orbit maintenance. Approximately 91 t of LH_2 is used during the TMI maneuver, including the "post-burn" cool-down propellant. The corresponding engine burn time is 39 minutes, which is well within the 62-minute single-burn duration that was demonstrated by the NRX-A6 engine during the NERVA program.

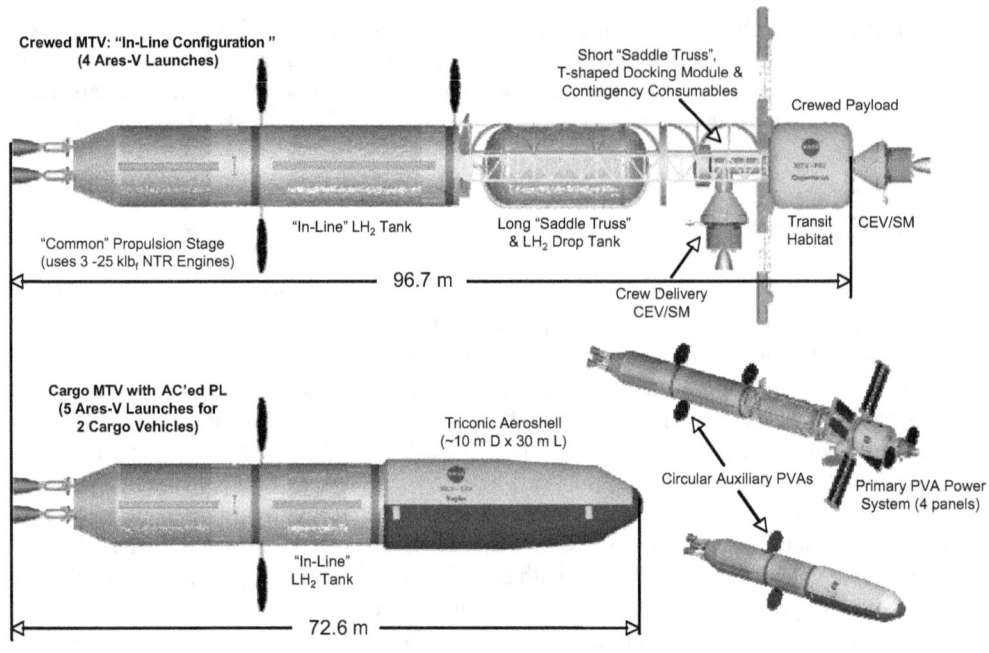

Figure 4-5. Crewed and cargo NTR design concepts.

The "all-propulsive" crewed MTV has an IMLEO of 356.4 t and an overall vehicle length of 96.7 m. It is an "in-line" configuration, which, like the cargo MTV, uses AR&D to simplify vehicle assembly. It uses the same common NTR propulsion stage but includes additional external radiation shielding on each engine for crew protection during engine operation. It also includes two saddle trusses that are open on the underside for jettisoning of the drained LH_2 drop tank and unused contingency consumables at the appropriate points in the mission. Four 12.5 kWe/125 m^2 rectangular PVAs, which are attached to the front end of the short saddle truss, provide the crewed MTV with 50 kWe of electrical power for crew life support, propellant tank ZBO cryocoolers and high-data-rate communications with Earth.

Four Ares V launches over 90 days are used to deliver the crew MTV vehicle components, which include: (1) the NTR "core" propulsion stage (106.2 t); (2) an in-line propellant tank (91.4 t); (3) a saddle truss and LH_2 drop tank (96 t); and (4) supporting payload (62.8 t). The payload component includes a short saddle truss that connects the transit habitat and long-lived Orion/service module (SM), which are used for vehicle-to-vehicle transfer and "end of mission" Earth entry, to the rest of the MTV. Also attached to the short saddle truss forward adaptor ring is a T-shaped docking module (DM) that connects the contingency consumables container with the transit habitat's rear hatch. More importantly, this second DM provides additional access to the MTV for the crew delivery CEV/SM.

The crewed MTV also carries 8 t of RCS propellant, which is split between the core stage and truss forward cylindrical adaptor ring. For the round-trip crewed mission, the required total usable LH_2 propellant loading is 191.7 t and the corresponding total engine burn duration is 84.5 minutes (57.8 minutes for TMI, 16 minutes for MOI and 10.7 minutes for TEI), which is well within the 2-hour accumulated engine burn time that was demonstrated on the XE engine during the NERVA program. Details on the cargo and crewed MTV component masses, launch sequence, and timeline are summarized in table 4-1.

Table 4-1. Reference NTR Manifest Summary

Reference NTR Transfer Vehicle Summary

Cargo Mission (Single Vehicle, 1 of 2)			Crewed Mission		
Vehicle Elements		Mass (t)	Vehicle Elements		Mass (t)
NTR "Core" Stage	Core Stage Dry Mass	33.7	NTR "Core" Stage w/Ext. Rad. Shield	Core Stage Dry Mass	41.7
	LH2 Propellant Load	59.4		LH2 Propellant Load	59.7
	RCS Propellant Load	3.6		RCS Propellant Load	4.9
	Total Core Stage Mass	96.6		Total Core Stage Mass	106.2
	Number of Core Stage	1.0		Number of Core Stage	1.0
	Total Stage Mass	**96.6**		**Total Stage Mass**	**106.2**
In-Line LH2 Tank	In-Line Tank Dry Mass	10.8	In-Line LH2 Tank	In-Line Tank Dry Mass	21.5
	LH2 Propellant Load	34.1		LH2 Propellant Load	69.9
	RCS Propellant Load	1.7		RCS Propellant Load	
	Total In-Line Mass	46.6		Total In-Line Mass	91.4
	Number of Tanks	1.0		Number of Tanks	1.0
	Total In-Line Mass	**46.6**		**Total In-Line Mass**	**91.4**
			Long Saddle Truss & LH2 Drop Tank	Saddle Truss Mass	8.9
				Drop Tank Dry Mass	14.0
				LH2 Propellant Load	73.1
				Total Assembly Mass	**96.0**
			Payload Elements	Short Saddle Truss	4.7
				Conting. Food Canister	9.8
				2nd Docking Module	1.8
				Fwd RCS Prop Load	3.2
				Transit Habitat	32.8
				CEV/SM + Crew	10.6
Payload	Total Cargo Lander (Aeroshell, PL & Lander)	103.0		**Total Payload Mass**	**62.8**
Total Cargo Vehicle Mass		**246.2**	**Total Crewed Vehicle Mass**		**356.4**

Vehicle Assembly Timelines & ETO Delivery Manifest

	Launch Number	Launch Time Before TMI (days)	Launch Manifest	Shroud Length (m)	Launch Mass (t)
Cargo Mission (Two Vehicles)	Ares V Launches				
	1	-180	NTR TMI Core Stage 1	30.0	96.6
	2	-150	NTR TMI Core Stage 2	30.0	96.6
	3	-120	Twin In-Line LH2 Tank	30.0	93.2
	4	-90	Payload 1 (Cargo Lander)	30.0	103.0
	5	-60	Payload 2 (Hab Lander)	30.0	103.0
		-60	TMI Window Allowance		
			Total MTV Mass Delivered to Orbit		**492.3**
Crewed Mission	1	-150	NTR Core Stage	30.00	106.2
	2	-120	In-Line LH2 Tank	30.00	91.4
	3	-90	Truss & Drop Tank	30.00	96.0
	4	-60	Crew Payload Element	30.00	62.2
		-60	TMI Window Allowance		
	Ares I Launch (delivers astronauts to orbiting crew MTV)				
	1	-5	6 Mars Crew	n/a	0.6
			Total MTV Mass Delivered to Orbit		**356.4**
Ares V launches:	**9**			**Total IMLEO (t):**	**848.7**

4.4 In-Space Transportation: Chemical/Aerocapture Option

The chemical/aerocapture MTV vehicle concept option for this study was made up of multiple-stage vehicles consisting of separate propulsive elements for each major mission maneuver. The vehicle elements were designed to allow maximum design commonality, efficient Earth-to-orbit delivery, and efficient assembly in LEO. The mission architectures that were considered in this study use two cargo vehicles and one crew vehicle for each Mars mission, as shown in figure 4-6.

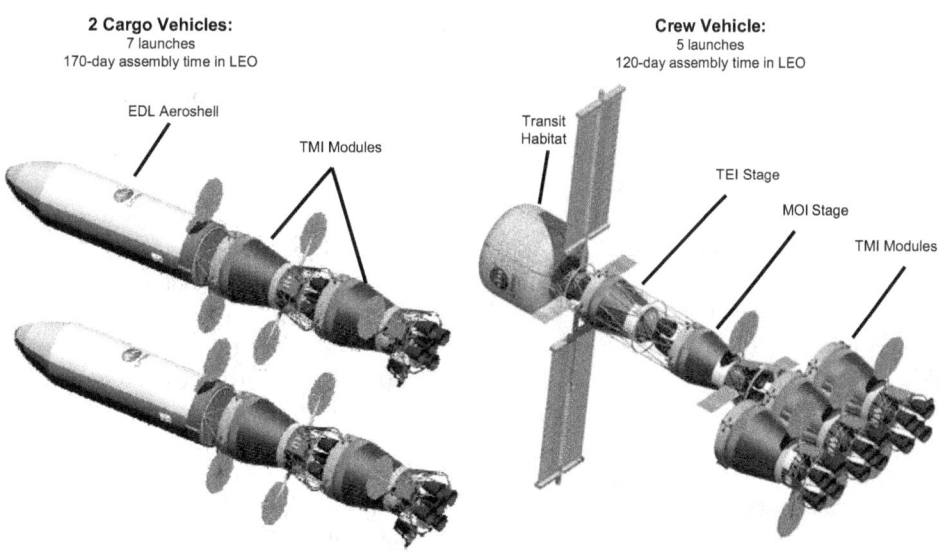

Figure 4-6. Chemical/aerobrake cargo and crewed MTV concepts.

The cargo vehicles would depart Earth approximately 2 years before the crew vehicle. One cargo vehicle would transport the Mars SHAB as payload, and the other would transport the Mars DAV as payload. The cargo vehicles consist of a payload that is enclosed in a cylindrical aeroshell and propulsive stages for TMI. The aeroshell would serve as a payload shroud for Earth-to-orbit launch of the payloads and an aerodynamic lifting body for Mars aerocapture, entry, and descent. Depending on the specific trajectory case, two or three TMI modules are required for each cargo vehicle.

The crewed vehicle consists of the CEV, transit habitat, three TMI propulsion modules, one MOI propulsion module, and one TEI propulsion module. The trans-Mars injection maneuver is divided into two propulsive burns. The two outboard TMI modules perform the first burn and are then jettisoned. The center TMI module performs the second burn. The CEV is used to transport the crew to LEO prior to TMI. A separate block upgrade version of the Orion vehicle remains docked to the transit habitat until shortly before Earth return, when the crew would separate from the transit habitat and perform a direct-entry Earth return.

Each MTV vehicle in LEO requires a LEO assembly reboost module, which performs attitude control and orbital reboost of the MTV during the LEO periods. The reboost modules are jettisoned from the vehicle stack prior to TMI.

4.4.1 Trans-Mars injection module
All of the TMI modules are common among the cargo and crew vehicles. Each vehicle performs a two-burn departure; since the required propellant for the crew mission drove all but one design, in the case of the cargo vehicles and one crew vehicle some propellant was offloaded. Each TMI module is jettisoned after it performs its burn. In the baseline cases, five RL10-B2 engines were used; the avionics package on each TMI module provided independent guidance navigation and control until it was docked to the entire stack. However, each module is responsible for fulfilling its own power requirements through launch and while connected to the rest of the vehicle for the duration of the LEO loiter period. Launch loads were considered to size the structural members, and mostly metallic materials were used for the beams and propellant tanks. Some composite materials were sized for panels. All of the thermal components for the cryogenic propellant tanks were sized assuming that the vehicle longitudinal axis maintained a sun-pointing orientation during loiter in a 407-km circular LEO.

4.4.2 Mars orbit insertion modules (cargo and crew missions)
For the all-propulsive crew missions, the MOI modules use two RL10-B2 engines to complete the MOI burn. For the crew vehicle, the MOI stage provides independent guidance navigation and control until it is docked to the stack, the TEI module would then take over control of the vehicle stack. In the case of the cargo vehicle, where there is no TEI stage, the MOI module provides guidance, navigation, and control (GN&C) for the stack throughout the entire mission. It also supplies the command and data links to all other stages, the high-gain antenna (HGA) communications to Earth, and the lander payload data link and standby power. Again, mostly metallic materials were assumed and the thermal approach was the same as compared to the TMI modules.

4.4.3 Trans-Earth injection module
The TEI stage for the crew mission was a scaled-down version of the crew MOI module with a few additional components. It uses two RL10-B2 engines and performs the TEI burn as well as the plane change while in Mars orbit. It is responsible for providing the following: GN&C for stack throughout mission, the command and data links to all other stages, the HGA communications to Earth, and the transit habitat data link and power. Structure and thermal assumptions were the same as compared to the TMI module.

4.4.4 Low-Earth orbit assembly reboost module
For this study, a generic re-boost module was designed to perform all on-orbit station-keeping operations during the LEO loiter phase of the mission. The re-boost module was required due to the long on-orbit loiter time and the mass of the vehicle that was being assembled. With a 407-km assembly and departure orbit, the vehicle is subjected to a small amount of drag that could decrease its orbit altitude over time. As with the ISS, the vehicle that is being assembled on orbit would periodically need to be re-boosted so the rendezvous and docking altitude remains constant.

The results of the final chemical/aerocapture architecture and system sizing are provided in table 4-2. As can be seen from this table, the reference chemical architecture requires 12 Ares V launches for each human mission to Mars. This reference approach maximizes the commonality between the Mars propulsion modules and the Ares V launch system to the greatest extent possible.

Table 4-2. Reference Chemical/Aerocapture Manifest Summary

Reference Chemical / Aerocapture Vehicle Summary

Cargo Missions (Both Vehicles)			Crew Mission		
Veh. Element		Mass (t)	Veh. Element		Mass (t)
TMI Stage 1	Mbo (Module)	15.1	TMI Stage 1	Mbo (Module)	15.1
	M prop (Module)	86.2		M prop (Module)	91.1
	RCS (Module)	2.3		RCS (Module)	2.3
	Total Module Mass	103.6		Total Module Mass	108.5
	Number of Modules	1.0		Number of Modules	2.0
	Total Stage Mass	**103.6**		**Total Stage Mass**	**217.0**
TMI Stage 2	Mbo	15.1	TMI Stage 2	Mbo	15.1
	M prop	86.2		M prop	91.1
	RCS (Module)	2.2		RCS (Module)	2.3
	Total Module Mass	103.6		Total Module Mass	108.5
	Number of Modules	1.0		Number of Modules	1.0
	Total Stage Mass	**103.6**		**Total Stage Mass**	**108.5**
			MOI Stage	Mbo	10.3
				M prop	50.2
				RCS	5.3
				Total Stage Mass	**65.8**
			TEI Stage	Mbo	11.4
				M prop	24.1
				RCS	7.3
				Total Stage Mass	**42.7**
Paylaod	Surface hab	103.0	Paylaod	Transit Habitat	41.3
				CM + Crew	10.6
Total Vehicle Mass		**310.2**	**Total Vehicle Mass**		**486.0**

Vehicle Assembly Timeline & ETO Delivery Manifest

	Launch Number	Launch Time Before TMI (days)	Launch Manifest	Shroud Length (m)	Launch Mass (t)
Cargo Mission (Both Vehicles)	Ares V Launches				
	1	-270	Reboost Module 1 Reboost Module 2	14.00	96.9
	2	-240	Payload 1 (Surf. Hab)	30.00	103.0
	3	-210	Payload 2 (Lander)	30.00	103.0
	4	-180	TMI Module 1a	16.26	103.6
	5	-150	TMI Module 2a	16.26	103.6
	6	-120	TMI Module 1b	16.26	103.6
	7	-90	TMI Module 2b	16.26	103.6
		-60	TMI Window		
	Total Mass Delivered to Orbit				**717.3**
Crew Mission	8	-210	Transit Hab/CEV Reboost Module	17.00	99.9
	9	-180	MOI & TEI Stages	22.30	108.5
	10	-150	TMI Module 1a	16.26	108.5
	11	-120	TMI Module 1b	16.26	108.5
	12	-90	TMI Module 1c	16.26	108.5
		-60	TMI Window		
	Ares I Launches				
	1	-5	6 Mars Crew	n/a	0.6
	Total Mass Delivered to Orbit				**534.5**
Ares V launches:	**12**			**Total IMLEO (mt):**	**1,251.8**

4.5 Launch Processing

There are significant ground and launch processing challenges for the elements and systems in support of human exploration of Mars. As was discussed previously, the number of launches necessary for human Mars missions varies depending on the final approach taken and is driven predominately by the type of in-space propulsion technology that is used. The total number of Ares V cargo launches for DRA 5.0 ranged from as low as seven to as high as 12 launches per human mission. Trades between the spacing of launches and subsequent impacts to the ground infrastructure and workforce as well as the total time required to launch the necessary mission elements, which directly impacts LEO loiter time and corresponding system reliability requirements, must be made. The ground operations assessment for the Mars campaign evaluated the infrastructure changes that are required to support 30-day launch centers, which was determined to be a good balance between the competing processing and the mission requirements. In addition, it was determined that between 2 and 6 months of schedule margin be included in the overall launch campaign to accommodate the launch processing, weather, and hardware anomalies the would most likely occur during the launch campaign.

Figure 4-7. Mars aeroshell and payload lifting operations in the VAB.

The Ares V launch vehicle ground operations concepts are still in the very early stages of development and, thus, the total impacts of both lunar and Mars mission needs must be evaluated together. Several Ares V vehicle concept trades are underway that would likely affect facility usage requirements as well as ground operations timelines. A "Ship to Integrate" ground processing concept was assumed for the Ares V core and some MTV elements (figure 4-7). The Ship to Integrate concept assumes that very limited ground processing activities are required at the launch site to prepare flight hardware for processing. The flight hardware is essentially unloaded from the transporter, inspected for damage, and stacked directly on the mobile launcher. No provisions are made for long-term storage of the element or for significant repair capabilities at the launch.

Lunar mission ground operations architecture development as well as several different options of facility improvements and workforce scheduling were considered for Mars DRA 5.0 development. For Constellation lunar missions, discrete event simulation models indicate that operations in the Vehicle Assembly Building (VAB) at KSC would likely be the limiting factor in determining minimum launch spacing and annual launch rates. Conflicts in the VAB transfer aisle during lifting operations as well as solid rocket booster (SRB) quantity distance issues eliminate some options from consideration. Based on the trades conducted, the Offline Stacking Facility (OSF) option was chosen to develop overall processing times for the Mars launch campaigns. The OSF is a new facility that is dedicated to stacking boosters for Ares I and Ares V. This facility would include two stacking bays with the capability to stack two boosters inside each bay. SRBs are stacked on a launch mount in the facility. Once the SRBs are stacked, the mobile launcher is moved into the OSF. The SRBs are translated onto the mobile launcher, and the mobile launcher is moved back to the VAB to continue the buildup and integration of the Ares V. This option does address SRB quantity-distance constraints in the VAB. For the NTR case, it was assumed that a new Nuclear Processing Facility (NPF) was required to process the core stages that contain nuclear material and require special processing considerations that were not planned in the lunar baseline. For both the chemical and nuclear cases, a new Hazardous Processing Facility is assumed to process spacecraft and MTV elements prior to integration with the launch vehicle in the VAB. The facility requirements are driven primarily by the size of the spacecraft elements and the desire to perform as much spacecraft processing as possible prior to critical path operations in the VAB and on the launch pad.

The ground processing concepts for this study focused primarily on launch vehicle facility and infrastructure impacts to meet 30-day launch spacing requirements. The results only provide very high level insights into changes that are required at the launch site to support the proposed Mars campaign above what is planned to support lunar missions, and further assessments are warranted.

4.6 Crew Exploration Vehicle/Earth Return Vehicle

Within the framework of the Mars DRA 5.0, a future block upgrade of the Orion CEV (figure 4-8) serves two vital functions: (1) the transfer of as many as six crew members between Earth and an MTV in LEO at the beginning of the Mars mission, and (2) the return of the as many as six crew members to Earth via direct entry from the Mars return trajectory. A CEV block upgrade (crew module and SM with a 3-year in-space certification) is launched as part of the crewed payload mass on an Ares V. The ISS version of the Orion, which will be launched by the Ares 1, delivers the six Mars crew members into an orbit that matches the inclination and altitude of the orbiting MTV. It then takes the CEV, which is conducting a standard ISS-type rendezvous and docking approach to the MTV, as many as 2 days to perform orbit-raising maneuvers to close on the MTV. After docking, the CEV, the crew performs a leak check, equalizes pressure with the MTV, and opens hatches. Once crew and cargo transfer activities are complete, the crew delivery CEV is jettisoned in preparation for TMI. The long-lived Orion block upgrade that was delivered on the Ares V is configured to a quiescent state and remains docked to the MTV for the trip to Mars and back to Earth. Periodic systems health checks and monitoring are performed by the ground and flight crew throughout the mission.

Figure 4-8. Orion crew exploration vehicle.

As the MTV approaches Earth upon completion of the 30-month round-trip mission, the crew performs a pre-undock health check of all entry-critical systems, transfers to the CEV, closes hatches, performs leak checks, and undocks from the MTV. The MTV is targeted for an Earth fly-by with subsequent disposal in heliocentric space. The CEV departs from the MTV 24 to 48 hours prior to Earth entry and conducts an on-board-targeted, ground-validated burn to target for the proper entry corridor; as entry approaches, the CEV CM maneuvers to the proper entry interface (EI) attitude for a direct-guided entry to the landing site. The CEV performs a nominal water landing, and the crew and vehicle are recovered. Earth entry speeds from a nominal Mars return trajectory may be as high as 12 km/s, as compared to 11 km/s for the lunar CEV. This difference will necessitate the development of a higher-density, lightweight TPS.

Two other factors (besides the primary concern of Earth entry speed) will drive the evolution of the CEV from a lunar vehicle to a Mars vehicle. The first is the need to re-certify the Orion for a 3-year on-orbit lifetime. Additionally, a science-driven mission to Mars would likely result in the desire to bring back an adequate amount of martian material (the current suggestion is 250 kg). Given the gear ratios involved in a round trip to Mars, the mass of such material would either have to be kept to a minimum or the upgrade would have to adopt an undetermined strategy by which to accommodate the mass and volume of this scientific material.

It was not within the scope of the DRA 5.0 activity to recommend specific design upgrades for the Orion vehicle or to develop an upgrade strategy. Instead, a mass estimate of 10 t was used for the vehicle CM to size propulsion stages. An additional 4 t was book-kept for a service module that may be needed to perform an Earth-targeting burn. Future activities, likely in conjunction with the Orion Project Office, will better define an upgrade strategy.

Depending on the trajectory flown, the entry speed of the Orion on a Mars return trajectory could be significantly higher than that for the lunar return at 11 km/s. Furthermore, since there would be a crew of six rather than four as would be the case for a lunar mission, the Mars block upgrade vehicle would be heavier than the lunar vehicle without incorporating other mass reduction efforts.

Figure 4-9 depicts the effect of g-constraints and vehicle mass on both peak heating rates and maximum heat load as a function of entry speed. The red dot at 11 km/s and mass of 9,227 kg corresponds to the Orion lunar vehicle with a heating rate of slightly less than 1,000 W/cm^2 and a heat load slightly more than 1,000 MJ/m^2. As can be seen in the figure, significant increases in heating rates and loads are introduced as the speed increases from 11 to 14 km/s while increasing maximum g's at a given entry speed results in less severe augmentations. Note that limiting the entry speed to the DRA 5.0 recommended limit of 12 km/s can provide significant reduction in TPS technology requirements as compared to previous studies with entry speeds up to 14 km/s.

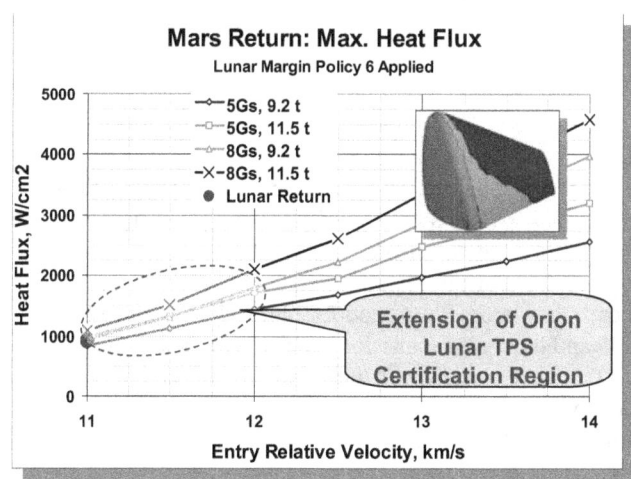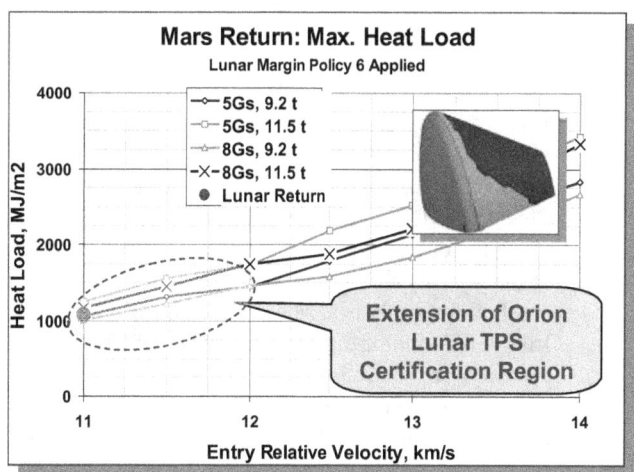

Figure 4-9. Orion TPS assessment results.

4.7 Mars Entry, Descent, and Landing

The baseline EDL system design was developed using the 10-m diameter × 30-m length dual-use launch shroud/entry aeroshell and a reference Mars orbit with a 1-sol period (250 km × 33,793 km). EDL system designs were developed for both of the cargo and habitat landers that use aerocapture for MOI while the crewed MTV uses propulsive MOI. In the case where aerocapture was used to achieve Mars orbit, the same aeroshell was used for both the aerocapture and the EDL phase, although additional TPS mass was required to accommodate the additional heating environment that is associated with the aerocapture maneuver. A pseudo-guidance methodology was developed to provide a realistic entry profile that would minimize terminal descent propulsive fuel requirements as well as the TPS mass and land the vehicle at 0 km Mars orbiter laser altimeter (MOLA) reference altitude. Several EDL configuration Architectures were considered during this study. They included an all-propulsive entry with no aeroassist elements, which was not selected because of the large orbit to landed payload mass fraction (on the order of eight) that was required for the payload masses that were considered. Supersonic aerodynamic decelerators, including parachutes and inflatable aerodynamic devices, were also considered for use in the descent phase, but the performance and mass models for the scale and dimensions that were required for the systems in this study were felt to be lacking in sufficient detail to be considered. The extrapolation in performance and masses from the references available were too large for these technologies to be weighed as viable options in the trade space. However, it is strongly recommended that future development of improved models for these types of systems technologies be pursued so that credible trades can be conducted and more optimal EDL system performance and reliability improvements realized. The reference EDL architecture that was ultimately selected for this study was a hypersonic aeroassist entry system, with a mid lift-to-drag ratio (L/D) aeroshell that was ejected at low supersonic Mach numbers. An LOX/liquid methane (LCH_4)-fueled propulsion system was used for deorbit delta-V maneuvers, RCS control during the entry phase, and final terminal descent to the surface.

The aerocapture and entry aeroshell structure mass estimates were made using preliminary estimates and guidance from the Ares V launch vehicle shroud development efforts. A "dual-use" launch shroud/EDL system approach was used whereby the launch vehicle shroud is used as the EDL structural element. The ellipsled/biconic shape was used to establish initial system performance characteristics, although other shapes, such as tri-conic systems, should be addressed in future studies. Aerocapture and EDL aeroshell structural mass estimates were based on equivalent area Ares V payload shroud mass sizing plus a 50% margin to allow for the additional lateral loads that are associated with entry and descent, TPS attachment scar mass, heat soak back, etc. The TPS analysis trade studies and sizing were conducted by personnel involved in the Orion – Crew Exploration Vehicle Thermal Protection System Advanced Development Project. The aerothermal environments that were associated with the Mars aerocapture and entry trajectories were determined using the NASA CBAERO tool, which was modified for use in the Mars atmosphere. The database for the CBAERO code was developed from a sparse set of high-fidelity, real-gas computational fluid dynamics (CFD) solutions from the DPLR code, combined with the line-by-line radiative heating code NEQAIR to provide predictions of convective and radiative heating solutions. Each solution contained full surface aerothermal environments including surface pressure, temperature, shear and uncoupled, convective, and radiative heating. While these codes represent the current state of the art, much uncertainty remains in understanding of large-scale entry vehicle environments for Mars. To account for uncertainties in aerothermal environments, margins were applied. Figure 4-10 depicts the distributions of mission maximum surface heating rates over the point design vehicle. The image to the left shows the distribution during the aerocapture phase, while that to the right shows the distribution during the out-of-orbit (1 sol) entry. The heating distributions during the out-of-orbit entry phase are independent of the method of orbital insertion, either by aerocapture or by propulsive means. The TPS materials selected for the aeroshell forebody heat shield were phenolic impregnated carbon ablator (PICA) and LI 2200. PICA is a candidate Orion/CEV ablator that is being developed for both the LEO and lunar return missions. PICA was the required TPS to account for the relatively high heating rates (462 W/cm^2) that were experienced during the aerocapture phase. For the leeward surfaces that are exposed to less severe thermal environments, heritage shuttle TPS materials were selected including LI-900 and felt reusable surface insulation (FRSI) blankets.

The descent stage engines were assumed, from previous large lander studies, to be RL10 derivatives and further assumed a thrust-to-weight ratio of the engines of 40 lbf/lbm. Recognizing that the LOX/LH_2 RL10 may not be the most appropriate analog for the LOX/LCH_4 engines that are currently baselined in this architecture, the parametric space was expanded to include engines that are derived from an RD-180 derivative that has a thrust-to-mass ratio of 80 lbf/lbm. The mass of the engines that were used in the thrust-to-mass ratio includes all associated turbopumps and all hardware that are attached to the engine before installation, but do not include the pressurant or tank-to-engine transfer line masses that were book-kept separately.

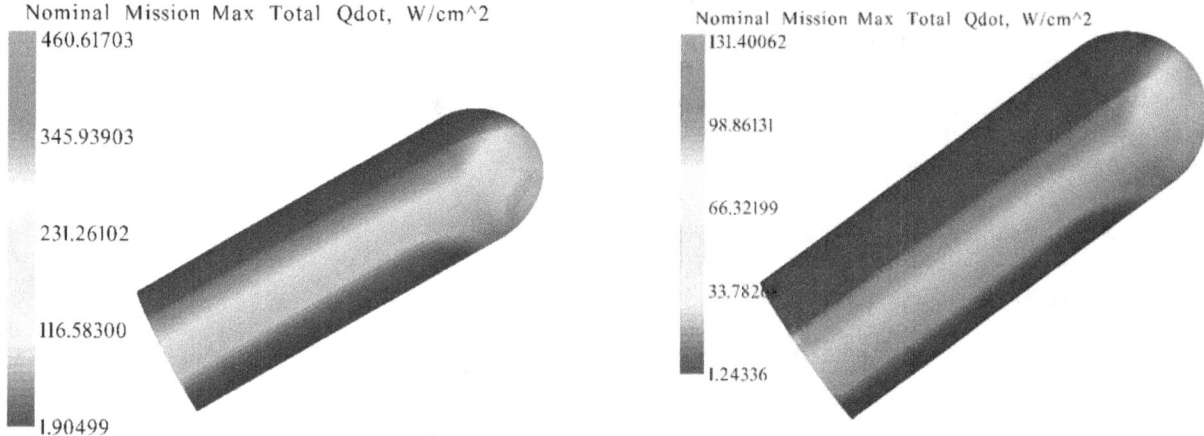

Figure 4-10. Distributed heating rates for the aerocapture and entry phases.

The descent stage dry mass is based on mass characteristics that were modeled using the Johnson Space Center (JSC) Envision mass sizing and simulation program. The descent stage is an all-propulsive, legged lander concept that uses four pump-fed LOX/LCH$_4$ engines with the following reference characteristics: an Isp of 369 sec, engine oxidizer-to-fuel (O/F) ratio of 3.5, chamber pressure of 600 pounds per square inch (psi), and a nozzle area ratio of 200. The baseline vehicle was sized to conform to the 10.0-m-diameter aeroshell. The descent stage thrust structure was assumed to undergo maximum loading during the descent maneuver and is sized to withstand the user-defined system thrust-to-weight ratio without the aeroshell attached as payload, assuming that the aeroshell was deployed prior to terminal descent engine initiation. In addition, the tanks of the descent stage are sized to include the deorbit fuel. Additional margin was place on the terminal descent fuel budget to perform a "divert maneuver" following the heatshield ejection so that the heatshield debris does not impact the surface near any highly valued pre-deployed assets. The final performance characteristics of the EDL system are provided in table 4-3.

Table 4-3. EDL System Characteristics

EDL Mass Summary			EDL System Characteristics		
Orbit Mass	**110.2**	t	Deorbit Delta-v	15	m/s
Deorbit Propellant	0.5	t	Balistic Coefficient	471	kg/m2
Entry Mass	**109.7**	t	Descent Delta-v	595	m/s
Aeroshell Structure	22.5	t	Max Heat Rate	131	W/cm2
Thermal Protection System	18.2	t	Total Heat Load	172	MJ/m2
RCS Dry Mass	1.0	t	Altitude Engine Initiation	1,350	m
RCS Propellant	1.2	t	Mach @ Engine Initiation	2.29	
Terminal Descent Propellant	10.1	t	Time of Flight	486	sec
Landed Mass	**56.8**	t	Time at Constant g's	134	sec
Dry Descent Stage	16.4	t	Engine T/W	161	N/kg
Payload Mass	**40.4**	t			

The Mars ascent vehicle that was used for the DRA 5.0 reference studies nominally transports a crew of six between the surface of Mars and the MTV (which has been loitering in Mars orbit for the duration of the surface mission). In light of the recommendation that ISRU technologies are used for ascent oxidizer production, the Mars ascent vehicle is pre-deployed to the surface of Mars during the opportunity prior to the crew's departure. It is only after the ascent vehicle is verified as fully fueled that the crew is committed on its journey via the MTV. For the

DRA 5.0 design activities, the JSC Envision parametric tool was used to size the lander, which was very similar (at a subsystem level) to the lander that was used in the Dual Lander Study (1999). This earlier vehicle, however, consisted of a two-stage ascent vehicle design with one LOX/LCH$_4$ pump-fed engine on the second stage and four on the first and descent stage. The LCH$_4$ fuel is brought from Earth, but the LOX is created using ISRU technology. This allows for a significantly lighter landed mass that propagates back through the architecture to result in substantially reduced IMLEO. This analysis resulted in a total ascent stage mass of approximately 21.5 t, including the necessary LCH$_4$ brought from Earth. The ascent engine characteristics of the LOX/LCH$_4$ pump-fed engines used 30-klbf engines running at 900-psi chamber pressure with a nozzle area ratio of 200.

4.8 Mars Transit Habitat

The crewed MTV consists of propulsion stages and propellant tanks for the TMI, MOI, and TEI maneuvers for both the nuclear or chemical propulsion options; the CEV that serves the function of an ERV for the final leg of the journey home; and a transit habitat in which the crew lives for the round trip between Earth and Mars. Although whether the transit habitat is constructed using rigid body or inflatable technology will need to be determined by detailed engineering analysis, it is assumed that it will share as many systems as pragmatically possible with the Mars SHAB. The rationale behind maximizing the commonality between these two elements (one that operates in a zero-g environment and the other that operates in a 1/3-g environment) is driven by the desire to lower the development costs as well as to reduce the number of systems that astronauts would have to learn to operate and repair. An even more critical assumption is that the systems comprising the transit habitat (and SHAB) would be largely based on hardware design and reliability experience gained by ISS operations, as well as long-duration surface habitat operations on the lunar surface (i.e., lunar outpost), which would precede any Mars campaign.

The mass estimates for the transit habitat are similar to the estimates that were used in DRM 4.0, but include a few changes in assumptions regarding dry weight margin (doubled to 30%) and the addition of spares for needed maintenance of the habitat.

A summary of these estimates is included in table 4-4.

Table 4-4. Transit Habitat Mass Summary

Transit Habitat Mass Estimate	Mass (kg)	Stowed Vol. (M3)
1.0 Power System	5,840	-
2.0 Avionics	290	0.1
3.0 Environmental Control & Life Support	3,950	19.1
4.0 Thermal Management System	1,260	5.3
5.0 Crew Accommodations	4,210	29.7
6.0 EVA Systems	870	2.9
7.0 Structure	2,020	-
Margin (30%)	4,920	8.6
Additional Spares	4,180	1.4
Crew	560	-
Total Transit Habitat Mass (without food)	**28,100**	**65.8**
Food (Return Trip)	2,650	7.9
Food (Outbound Trip)	2,650	7.9
Food (Contingency)	7,940	23.5
Total Consumable Mass	**13,240**	**39.4**
Total TransHab Mass @ TMI	**41,340**	
Crew Exploration Vehicle (assumed mass)	10,000	
Total MTV Payload Mass @ TMI	**51,340**	

The food that is carried aboard the transit habitat includes transit consumables that are needed for the round-trip journey plus contingency consumables that are required to maintain the crew should all or part of the surface mission be aborted and the crew forced to return to the orbiting MTV, which would then function as an orbital "safe haven" until the TEI window opens. Any remaining contingency food remaining on board the crewed MTV would be jettisoned prior to the TEI burn to return home.

5 SURFACE SYSTEMS

Technical studies that are associated with the surface systems for DRA 5.0 consisted primarily of understanding the relationship between the functional capabilities that are necessary to accomplish the exploration goals and objectives and the establishment of the top-level definition of the systems that are necessary for those functions. In most cases, detailed designs were not developed but, rather, top-level performance estimates and trades were conducted. More in-depth detailed definition of the various surface systems should be conducted in future efforts, including commonality with lunar surface systems.

Advanced Exploration – An artist's concept of subsurface access achieved by drilling on Mars. Frassanito 2003.

Deliberations by the science team determined that surface mobility, including exploration at great distances from the landing site as well as subsurface access, were keys to a robust science program. To understand the implications of these goals on the resulting surface systems, a range of surface strategies were considered, each of which emphasized a differing mix of mobility, depth of exploration, and duration of exploration in the field. These surface strategies included: (1) **Mobile Home**: emphasizing long-duration exploration at great distances from the landing site via the use of large, pressurized rovers; (2) **Commuter**: providing a balance of habitation and small pressurized rovers for mobility and science; and 3) **Telecommuter**: emphasizing robotic exploration enabled by teleoperation of small robotic systems from a local habitat. Each of these scenarios was used to provide a better understanding of the systems and capabilities that are needed to accomplish surface exploration goals.

The "**Commuter**" surface mission scenario was adopted as the nominal scenario for this reference architecture. This scenario included a centrally located, monolithic habitat, two small pressurized rovers, and two unpressurized rovers (roughly equivalent to the Apollo LRV). Power for these systems would be supplied by a nuclear power plant that would be previously deployed with the decent-ascent vehicle and used to make a portion of the ascent propellant and consumables (H_2O, oxygen (O_2), and buffer gases) to be used by the crew when they arrive. Although traverses would be a significant feature of the exploration strategy that is used in this scenario, these would be constrained by the capabilities of the small pressurized rover. In this scenario, these rovers have been assumed to have a modest capability, notionally a crew of two, 100 km total distance before being re-supplied, and no more than 1 week duration. Thus, on-board habitation capabilities would be minimal in these rovers.

With the limited resources that were available for this study, a very preliminary estimate was made of the mass for each of the surface system elements and their distribution between the two cargo elements that would be used to deliver them to Mars. Table 5-1 provides a summary of these payload masses and their distribution between the two landers.

Table 5-1. Mass Summary for the "Commuter" Surface Scenario

Surface Systems	Quantity	Habitat Lander System Mass (kg)	DAV Lander System Mass (kg)
Crew Consumables	-	1,500	4,500
Science	-	-	1,000
Robotic Rovers	2	-	500
Drill	1	-	1,000
Unpressurzed Rover	2	-	500
Pressurized Rover	2	8,000	-
Pressurized Rover Growth	-	1,600	-
Pressurzed Rover Power	2	-	1,000
Traverse Cache	-	-	1,000
Habitat	1	16,500	-
Habitat Growth	-	5,000	-
Stationary Power System	2	7,800	7,800
ISRU Plant	2	-	1,130
Total Surface Systems	-	**40,400**	**18,430**

Lander Systems	Quantity	Habitat Lander System Mass (kg)	DAV Lander System Mass (kg)
Ascent Stage 1 (no LOX)	1	-	12,160
Ascent Stage 2 (no LOX)	1	-	9,330
Descent Stage (wet)	2	23,760	23,760
Aeroshell	2	42,900	42,900
Total Wet Mass (IMLEO)	-	**107,060**	**106,580**

5.1 Surface Habitation Systems

Development of the Mars DRA 5.0 was conducted at the same time that formulation of various lunar surface scenarios was being conducted by the LAT. One of the key strategies of the lunar missions is the development and demonstration of fundamental exploration capabilities that could be used for future exploration beyond LEO; i.e., Mars. Due to time and resource limitations, a detailed assessment of Mars habitats was not conducted. Instead, emphasis was placed on understanding the fundamental similarities and differences between the lunar and Mars habitation systems. The first step in the approach was to establish ground rules and assumptions. This defined the excursion range, crew size, and other attributes for each of the options. Next, a master equipment list (MEL), which was created for recent lunar habitat studies, was used as a point of departure for the Mars options (figure 5-1). This was a logical starting point because space habitats share similar subsystems, and the MEL incorporated the latest detailed input from subsystem specialists. Each of the subsystems was examined to determine the mass and power changes that are required to accommodate the Mars habitat options. The reference approach for DRA 5.0, the Commuter option, had a habitable base that remained on the lander and used two small pressurized rovers for exploration excursions.

Lunar habitats accommodated a crew of four and varied from an assembly of small modules to a one-shot delivery to a "train" of smaller mobile homes. Modifications were necessary for crew size, overall mission duration, and logistics capabilities. Due to limited opportunity for logistics resupply for Mars missions, each subsystem determined a spares factor of additional mass to be delivered with the habitat. For totals, a 20% concept design factor was added. The Commuter habitat approach is approximately 21.5 t using 12.1 kWe of electrical power.

A key objective of the Mars surface mission is to get members of the crew into the field where they could interact as directly as possible with the planet that they have come to explore. This would be accomplished via the use of EVAs, assisted by pressurized and unpressurized rovers, to carry out field work in the vicinity of the surface base.

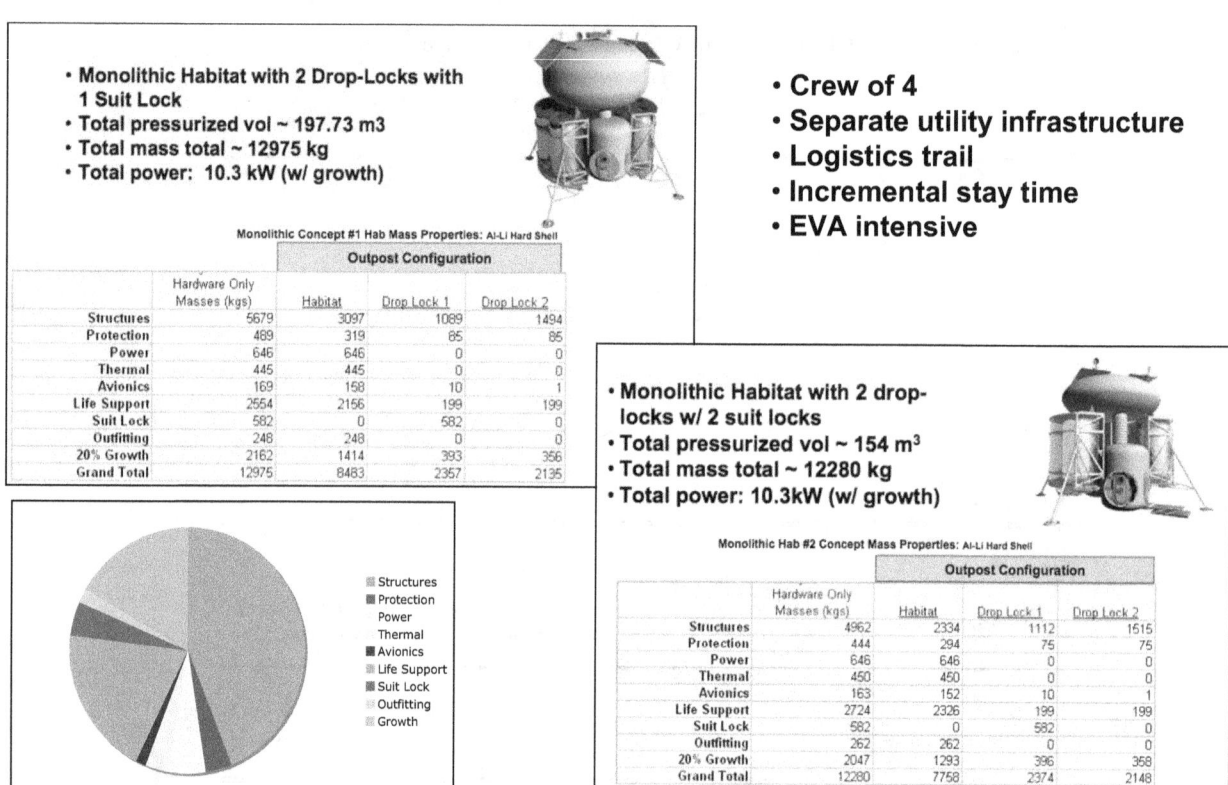

Figure 5-1. Mars habitats draw from lunar architecture options.

5.2 Surface Mobility Systems

A typical field exploration campaign would begin with one or more questions regarding the geology in a particular region and the identification of specific surface features, which are based on maps and overhead photos that offer the potential for answering these questions. Traverses are planned to visit these sites, typically grouping these sites together (into multiple traverses, if necessary) to meet the limitation of the equipment or environment (e.g., EVA suit duration limits, fueled rover range, crew constraints, local sunset, etc.). Depending on the anticipated difficulty of the planned traverse, the crew may choose to send a teleoperated robot to scout the route that would send back imagery or other data for the crew to consider. In addition, crew safety concerns when entering a region that is highly dissimilar from any explored before or an area with a high potential for biological activity may dictate the use of a rover in advance of the crew.

Several key scientific and operational questions would require subsurface samples that are acquired by drilling. Examples include searching for subsurface H_2O or ice, obtaining a stratigraphic record of sediments or layered rocks, or obtaining samples to be used to conduct a search for evidence of past or extant (possibly endolithic) life. Drill equipment would be moved to the site, most likely on a trailer that is pulled by either the unpressurized or robotic rovers, and set up for operations. The set-up process would likely be automated, but with the potential for intervention by the crew. Drilling operations are also likely to be automated but under close supervision by the crew. At present, drilling is still something of an art, requiring an understanding of both the nature of the material being drilled – or at least a best guess of the nature of that material – and the equipment being used. While drilling is a candidate for a high level of automation, it is likely that human supervision for purposes of "fine tuning" the operations and intervening to stop drilling would remain a hallmark of this activity. Core samples would be retrieved by the crew and put through an appropriate curation process before eventual analysis. After concluding drilling at a particular site, the drill equipment would be disassembled and moved to the next site, where this procedure would be repeated.

Because of the nature of the drilling process, it is highly probable that the above-surface equipment would fail or the below-surface equipment would break or seize. Crew intervention is highly likely in either event. In the first case, the crew must decide whether the failure could be fixed in the field or whether the equipment must be returned to the outpost for repair. Due to cargo mass constraints, the drill would not have an unlimited supply of drill bits, auger bits, or drill stem. This makes it worthwhile to expend some effort to retrieve as much of the salvageable subsurface equipment as possible and attempt a repair – the alternative being to halt drilling operations until adequate replacements arrive, probably with the cargo flights supporting the next crew.

As is apparent in the previous discussion, conducting geologic investigations on the surface of Mars would require extensive EVA to take advantage of the human element over robotic rovers. The EVA system, therefore, is a critical element in maximizing the science return from a human Mars mission. The EVA system that is currently under development for the lunar surface would require modifications to operate under environmental conditions on Mars. Three characteristics of the martian environment dictate this: (1) increased value of the surface gravity from 1/6 g on the lunar surface to 1/3 g on Mars; (2) the change in the atmosphere from essentially vacuum to an approximately 10 mbar CO_2 and argon (Ar) atmosphere; and (3) the requirement to minimize contamination of the martian environment and exposure of astronauts to martian materials.

5.2.1 Surface transportation: unpressurized and pressurized rovers

Even at distances that are considered within walking range, incorporation of surface transportation has been found to enhance crew productivity, both to mitigate crew fatigue and to extend consumable supplies by allowing lower metabolic rates during seated travel. Providing the capability to travel easily and quickly away from the landing site would be necessary for the crew to remain fully productive throughout the surface mission.

The unpressurized rover could be viewed in many ways as an extension of the EVA suit. From this perspective, many of the heavier or bulky systems that would otherwise be an integral part of the suit could be removed and placed on the rover, or the functionality of certain systems could be split between suit and rover. In the case of offloading capabilities to the rover, navigation, long-range communication, tools, and experiment packages could be integrated with or carried by the rover. In the case of splitting functionality, any of the various life support system consumables (e.g., power, breathing gases, thermal control, etc.) could be located on both the rover and within the EVA suit. This division or reallocation of EVA support functionality may restrict the maximum duration of the EVA suit to something less than that which has been previously demonstrated. However, analysis of Apollo LRV exploration indicates that approximately 20% of the total EVA time was spent by the crew on the LRV moving from site to site. Mars surface operations could be assumed to be comparable. Thus the EVA team would have sufficient time for recharge of EVA suit consumables or switching to rover-based support systems to preserve EVA suit consumables. Providing multiple sources of consumables and support systems in the field also enhances crew safety by providing contingency options should EVA suit systems degrade or fail.

Operationally, Mars surface EVAs would be conducted by a minimum of two people and a maximum of four. (This would always provide for a "buddy system" while on an EVA but would also leave at least two people in the SHAB for contingency operations should they be needed.) If unpressurized rovers are used, an additional operational constraint would be imposed on the EVA team. If one rover is used, the EVA team would be constrained to operate within rescue range of the surface base. This could mean either the team has sufficient time to walk back to the surface base if the rover fails, or that there is sufficient time for a rescue team from the surface base to reach them. Taking multiple, and identical, rovers into the field allows the EVA team to expand its range of operation because these vehicles are now mutually supporting and, thus, able to handle a wider range of contingency situations. Operationally, the rovers must be reliable but also easily repairable in the field (or at least have the capability of being partially disassembled in the field so that the failed component could be returned to the outpost for repair). The rovers must also be sized to carry cargo that, if offloaded, is of a sufficient capacity to carry the crew of a disabled rover.

Pressurized rovers are typically included in the Mars mission studies because of their ability to extend the range of the crew, in terms of both distance and duration. While exact distances and durations would be dependent on the specific site chosen, input received from the HEM-SAG indicates a strong desire to reach locations several hundred kilometers from the outpost for durations measured in days to weeks between resupply. It was also the intent that the crew using the pressurized rover be capable of performing many of the same functions as at the outpost, albeit at a reduced scale. Thus a crew using a pressurized rover could be expected to be capable of commanding and controlling teleoperated rovers, conducting EVA activities (comparable to those discussed earlier) within the vicinity of the rover, and otherwise supporting the crew for the duration of its excursion away from the outpost.

For this DRA assessment, a modest pressurized rover capability was assumed. This rover was scaled to support a crew of two (with the ability to support four people in a contingency) for a period of approximately 2 weeks without resupply and travel for a total distance of approximately 100 km. These two pressurized rovers are assumed to be nimble enough to place the crew in close proximity to features of interest (i.e., close enough to view from inside the rover or within easy EVA walking distance of the rover).

5.3 In-Situ Resource Utilization

The ISRU plant is designed to convert Mars atmosphere into O_2 for use as propellants and life support. In addition to O_2, the ISRU system generates H_2O and buffer gases for use in the surface habitats and mobility systems. The plant is made up of a solid oxide CO_2 electrolyzers (SOCEs) that convert CO_2 into O_2 and carbon monoxide (CO,) which is vented. The CO_2 is obtained via a micro-channel adsorption pump. The CH_4 fuel that is required for ascent is brought from Earth. Hydrogen (H_2) (400 kg) is brought from Earth and reacted with Mars-produced O_2 to make up H_2O that is lost during crew and EVA operations. Besides CO_2, N_2 and Ar are also separated and collected from the Mars atmosphere for use as a buffer gas for crew breathing.

The atmospheric acquisition ISRU plant was modeled by dividing it into three subsystems: the atmospheric acquisition subsystem, the consumable generation subsystem, and the liquefaction subsystem. The atmospheric acquisition subsystem is made up of the following component models: filter, micro-channel CO_2 adsorption pump, valves, flow controllers, buffer gas pump, and buffer gas tank. The consumable generation subsystem is made up of an SOCE, heat exchanges, filters, and valves. The liquefaction subsystem is made up of cryocoolers for CH_4 and O_2, filters and valves. Since the plant is driven more by power than mass, redundancy is accomplished by the use of two separate ISRU plants, each sized to generate the needed consumables. The atmospheric acquisition ISRU plant model provides results for mass, power, and volume for each subsystem in the plant. The results were based on the above plant producing all of the necessary O_2 for an ascent vehicle as well as consumables for the Environmental Control and Life Support System (ECLSS), which consist of H_2O, O_2, and inert gases (N_2 and Ar) that are a byproduct of the Mars atmosphere. The mass, power, and volume of the system and associated components is recorded in table 5-2. These estimates are based on continuous propellant production, which is provided by a nuclear fission power source. Power estimates for a solar-based system are much higher since propellant production could only be done during the day, which requires a far greater processing rate and subsequent power level.

Table 5-2. In-Situ Resource Utilization System Mass

	Quantity	Unit Mass (kg)	Total Mass (kg)	Volume (m3)	Power (kWe)
Atmospheric Acquisition Subsystem	2	-	492.12	0.66	17.86
Filter/Frit	4	0.10	0.40	-	-
Microchannel CO2 Adsorption Pump	4	57.50	230.00	0.01	17.86
check Valve	8	0.10	0.80	-	-
Buffer gas pump	4	1.23	4.92	0.00	0.00
Isolation Valve	8	0.50	4.00	-	-
Buffer gas tank	1	250.00	250.00	0.60	-
Flow Controller	4	0.50	2.00	-	-
Oxygen Generation System	2	-	38.80	0.10	2.59
Solid Oxide Electrolysis Stack	2	17.00	34.00	0.05	2.59
Isolation Valve	8	0.50	4.00	-	-
Filter/Frit	4	0.10	0.40	-	-
check Valve	4	0.10	0.40	-	-
Liquefaction Subsystem	1	-	34.60	0.10	3.26
Hydrogen Cooler	2	10.60	21.20	0.01	0.34
Methane Cooler	2	1.20	2.40	0.01	0.02
Oxygen Crycooler	2	5.50	11.00	0.03	2.90
ISRU System (each)	**-**	**-**	**565.52**	**0.86**	**23.71**

5.4 Surface Power Systems
5.4.1 Stationary power

The reference stationary surface power generation system is a nuclear fission power reactor concept that is based on a lunar design. This lunar system was conceived to be easily adaptable to operation on the martian surface. The low operating temperature of the reactor fuel enables use of stainless steel for major reactor components, a material that is compatible with Mars' predominately CO_2 atmosphere. The nuclear power system's mass used for comparison was for a 30-kWe version of the 40-kWe lunar design to match the requirements of the Mars mission. The reactor would be landed in the DAV in a stowed configuration and offloaded from the cargo bay for emplacement using a utility power cart that would have multiple functions. The utility cart could be photovoltaic (PV) with regenerative fuel cells (RFCs) that are battery-powered or use a Radioisotope Power System (RPS). For this study, it was assumed that a Dynamic Isotope Power System (DIPS) would be used for the power cart and could also be an option for powering the pressurized rovers. The Plutonium 238 isotope, which has fueled numerous deep space missions as well the two Viking landers and long-term experiment packages on Apollo, would be used with advanced power conversion technology to increase power output from three- to four-fold when compared with thermoelectric devices that have been previously used. The advantage of this technology is that continuous power (24 hours a day/7 days a week) is Available from this unit without need for any recharging. It is envisioned that the DIPS cart would provide power to the reactor mobility chassis while it is being transferred to a location approximately 1 km from the landing site.

The primary surface reactor has an external shield to protect the crew from radiation. Similar to the lunar application, this study has adopted a guideline of less than 5 rem/yr dose to the crew from reactor-generated radiation. Since the shield is a significant portion of the system mass, a shaped shield is employed whereby the radiation is limited to 5 rem/yr (at 1 km) in the direction of the habitat and 50 rem/yr (at 1 km) in all other directions. This creates a small exclusion zone but still allows limited passage through the zone under special circumstances. Based on $1/R^2$, the radiation level in directions away from the habitat zone is just over 5 rem/yr at a distance of 3 km assuming a line of sight. One option to reduce or eliminate the exclusion zone and to save shield mass would be to bury the reactor below grade, where the soil provides additional protection, as has been suggested for lunar applications. However, the team felt that this option was risky due to numerous factors, such as the complexity of the remote operations required to bury the reactor, and has opted for the above-ground emplacement. If a second reactor were required for risk reduction, it would be possible to consider the crew assisting in burying and setting up the second nuclear power system, using power available from the first reactor unit.

With the above-ground option, the reactor would be driven about 1 km from the lander that is feeding out the power cable. Once at the site, the mobile chassis would be aligned to properly orient the shield, leveled, and secured by jacks. The DIPS cart, which would be outfitted with appropriate equipment, would assist in the deployment of the radiators if needed. The power cart would be driven back to the landing site and the reactor would be started. It was assumed that the total time to perform this is 30 to 40 sols.

Figure 5-2 shows the power that is required for the various architecture elements for normal day and night operations. The habitat power estimate is a modified lunar concept that has been scaled for Mars operations. Power systems were sized for a 12-kWe day/night load for the habitat when using the ISRU-produced O_2 supply. Additional habitat power would be required for closed-loop air revitalization. Further analysis of nighttime and dust storm habitat power needs is required to establish a minimum array area of the solar power system. The ISRU plant, which is making ascent stage O_2 propellant, is the dominant power requirement at 25 kWe operating continuously. After propellant production has been completed, most of the power demands are in support of nominal outpost operation, including habitats, logistics systems, rovers, scientific systems, and ascent stage keep-alive power. Thus, a power system that is sized to meet the ISRU consumable production requirements would have ample power available for crew outpost operations.

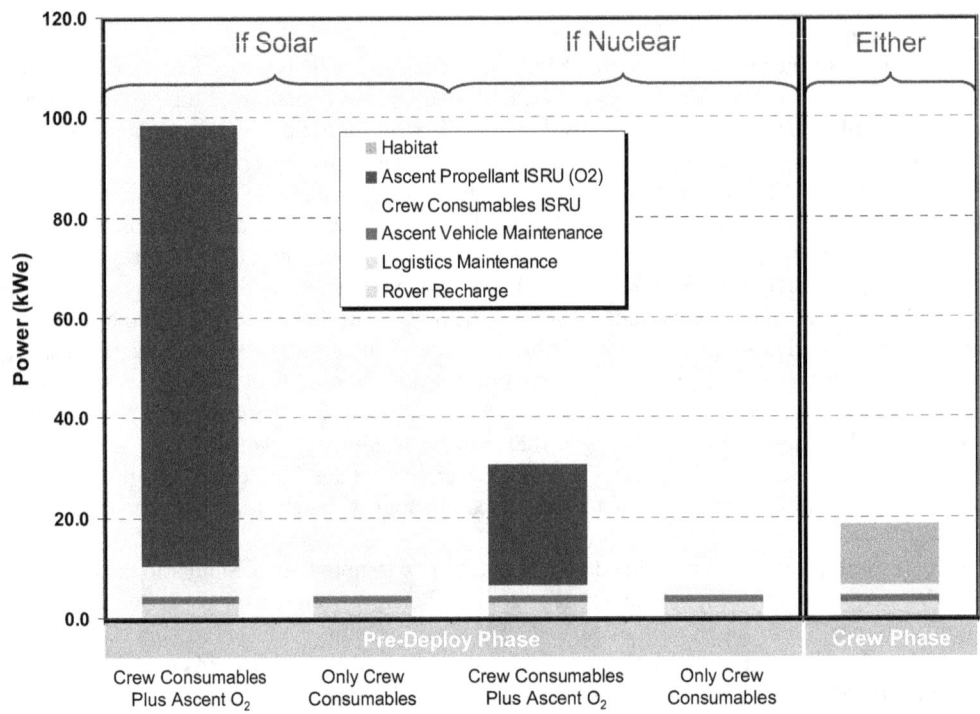

Figure 5-2. Power estimates for solar and nuclear power options.

5.4.2 Mobile power

In addition to the main base power system, options were looked at for powering the surface mobility systems (rovers). The reference "Commuter" strategy has two smaller rovers that would support a crew of two and traverse 100 km (130 km total with trafficability factors included) in 15 days. With this scenario, the central habitat is used by the crew between sorties. Since there were no operational details or timelines provided by the science team, a drive time was assumed to be 5 hours each day, which dictated a speed of 3 km/hr to cover the total distance in the time allocated (driving was only during sunlight). In addition, a "trafficability" factor of 30% (avoid rocks, steep grades, soft sand, etc.) was assumed to capture an "odometer" distance that rover speed would be based on, thus a total of 130 km is actually traversed during the sortie.

Three power system options were evaluated. These options included: PV/battery, PV/battery with DIPS augmentation, and fuel cell only (table 5-3). The significant drivers for both power and energy are the rover mass and drive speed. The drive power to achieve the 3 km/hr speed for the small rovers is approximately 25 kWe. This is a major challenge to meet the specified requirement of sortie distance in the allotted time. To keep the array area and battery mass to a minimum, it is necessary to recharge the system on as short a cycle as possible. Therefore, for this analysis, we adopted the operation scenario of driving and stopping to do science and recharge on alternating days. Even with this strategy, the array size that is required to recharge the batteries is 400 m^2, which must be deployed and stowed. If we assumed a 5-m-long rover and two 200-m^2 arrays, the crew would need to deploy each array approximately 40 m out from the rover. Adding a 2.5-kWe DIPS did not have much impact on the sizing due to the low ratio of load power to DIPS output at the 3-km speed. However, if the sortie's 15-day duration were relaxed, speed could be reduced and the resultant drive power reduced greatly. A speed of 0.5 km/hr brings the drive power close to the nominal crew power of 5 kWe. Array area and battery mass is reduced, and now the addition of the DIPS allows a major reduction in array area and battery mass. One additional case was evaluated at 0.1 km/hr to reduce the array area to a size that could be fixed on top of the rover, thereby eliminating need for array deployment/stowage. A fuel-cell-only option was assessed whereby the O_2 reactant could be produced by the ISRU plant during the pre-deploy phase. The O_2, H_2, and total fuel cell mass estimates are shown based on accomplishing the full sortie round-trip distance within the required duration.

Table 5-3. Small Pressurized Rover Power Options

Scenario	Speed (km/hr)	Array Area (m2)	Isotope Power (kWe)	Drive Power (kWe)	Battery Mass (kg)	Kwe-hr	DIPS Mass (kg)	Array Mass (kg)	Total Mass (kg)
Solar/Battery	3.0	400	-	25	2,500	250	-	1,000	3,500
Solar/Battery	0.5	160	-	4	1,100	116	-	400	1,500
Solar/Battery/DIPS	0.5	40	3	4	300	30	190	100	590
Solar/Battery/DIPS	0.1	10	3	1	130	13	190	25	345

Scenario	Speed (km/hr)	Fuel Cell System (kg)	ISRU Provided O2 (kg)	ISRU Provided H2 (kg)	Total Mass (kg)
All Fuel Cell	3.0	1,743	122	975	2,840

6 ARCHITECTURAL ASSESSMENTS

A primary focus of the DRA 5.0 development was the identification and systematic assessment of the principal key challenges that are associated with the human exploration of Mars. A top-down system engineering approach was established to identify, assess, and systematically eliminate unattractive options from further consideration. This process was facilitated by the development of an architecture trade tree (figure 6-1), which provides a graphical representation of the key technical linkages and architectural challenges that are associated with future human exploration missions to Mars. The trade tree was established to identify those key decisions or "architectural branches" that had the greatest overall leverage on the resulting architecture. Providing a structured approach allowed the study team to systematically eliminate complete branches, thus placing effort on those branches that provide the best balance of the key figures of merit (FOMs): safety, cost, and performance. The architecture trade tree was an effective tool allowing the team to strategically address the overall architectural approaches while concentrating on those options that provided the highest overall architectural leverage early in the study. The overall study approach was structured to begin with this high-level architectural "trade tree trimming" followed with a series of architecture refinement activities with the purpose of better optimizing the overall architectural approach.

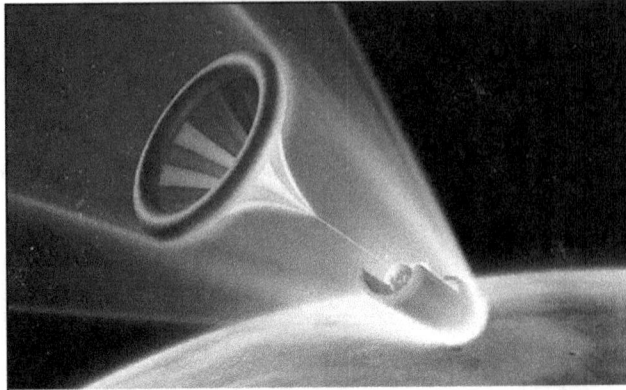

Final Descent – An artist's concept depicting one potential approach of achieving the difficult hypersonic deceleration maneuver during Mars entry. Rawlings 2007.

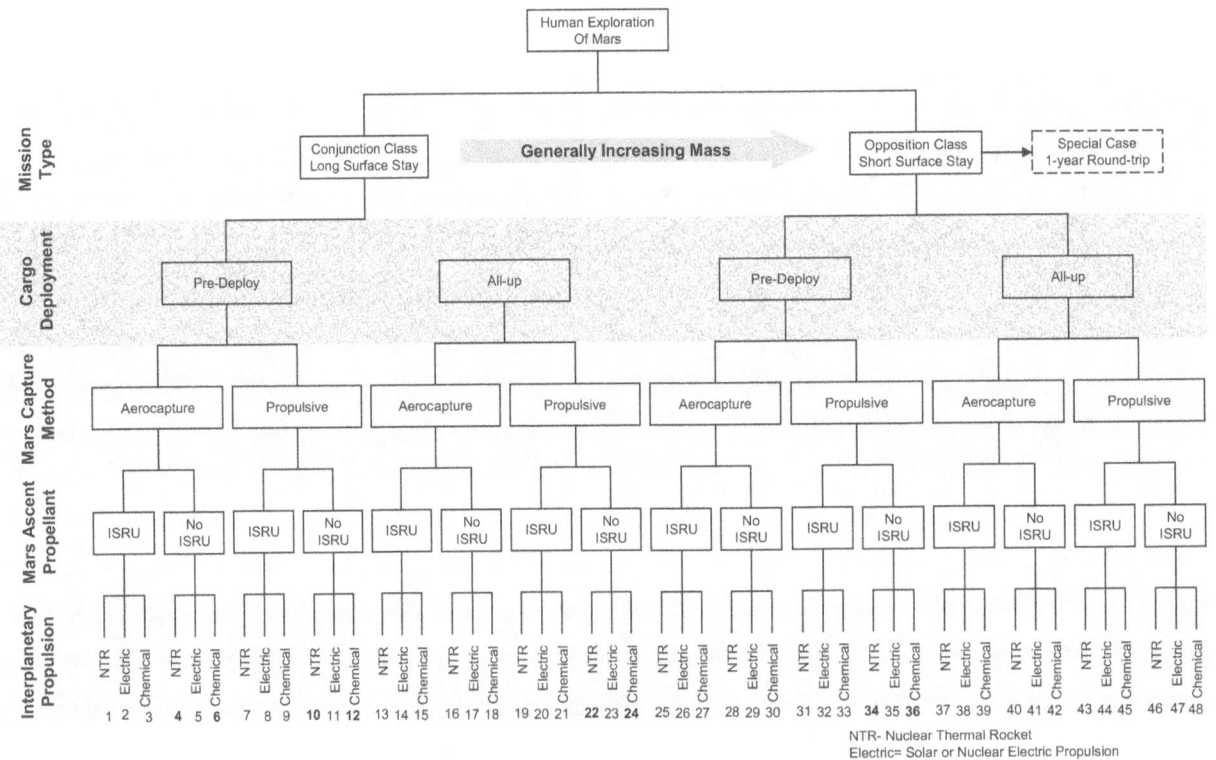

Figure 6-1. Mars Architecture Study top-level trade tree.

The emphasis of the first phase of the study activity focused on trimming the trade tree by developing specific decision packages that were associated with each key architectural branch of the trade tree. Each decision package used a common set of integrated performance tools that included an estimate of the overall architecture performance, risk, and cost. In addition, each decision package was formulated around a common set of FOMs with common key measures of effectiveness. Each decision package was then reviewed by the agency Joint Steering Group for concurrence on the results of the assessment. The Joint Steering Group was comprised of agency-wide leadership from the ESMD, SMD, ARMD, and SOMD as well as the Chief Health and Medical Officer. This iterative approach allowed an appropriate hierarchy of decisions to be addressed in a very systematic manner. Since the emphasis of this initial phase of assessments was on the key decision points of the trade tree, emphasis was placed on the relative comparison of the architectural approach that is associated with that specific decision comparison. That is, emphasis was placed on ensuring that a proper relative comparison between the two comparative branches was achieved as opposed to optimization of a specific branch. Emphasizing the relative architectural comparisons allowed the study team to develop rapid high-level comparative models rather than spending too much time refining the specific design details. Optimization was reserved for the second phase of the study within a narrower set of architectural options or branches. Emphasis during the first study phase was placed on establishing the proper level of details that is associated with the decision at hand to ensure that important design or operational details were not overlooked that could sway the decision in a different direction. To aid in this process, previous models and design details that were developed by various subject matter experts who participated in the many previous human exploration of Mars efforts were used to the greatest extent possible. Throughout this process, emphasis was placed on consistency and commonality of all ground rules, assumptions, and modeling approaches to ensure that the proper relative comparisons were being made.

During the development of DRA 5.0, emphasis was equally placed on the assessment of overall architectural risk and cost as well as performance. Integrated risk and cost models were developed that were based on the technical details that were developed by the various subject matter experts. These risk and cost concepts were then combined into an overall mission model for assessment of the overall architectural risk and cost. Assessment of the resulting integrated model allowed for the identification of the key cost and risk drivers that are associated with each represented branch of the architecture trade tree.

6.1 Figures of Merit

During the study, key FOMs were used to help the analysis team develop an understanding of implications of the various decisions under consideration. The FOMs were used to measure the benefit of one approach as compared to other alternatives. Using standard categories, consistent sets of measures made it possible to compare alternatives in addition to providing insight into the performance sensitivities of the alternatives and variations due to different assumptions and inputs. The specific measures of effectiveness that are associated with each FOM were established based on the specific decision on hand. The MAWG used the following FOMs in the development of the various decision packages under consideration.

6.1.1 Safety and mission success

Measures of effectiveness that are associated with safety and mission success focus on determining the degree to which a mission concept or technology option ensures safety and reliability for all mission phases. To be sustainable, future space exploration systems and infrastructure, and the missions that are pursued using them, must be reliable; and when astronauts are involved, they must be as safe as reasonably achievable. Emphasis is placed on understanding comparative values of the safety-related measures of performance that are discussed below:

- *Risks*. An assessment of the events that could result in loss of crew, loss of vehicle, and mission failure. These could include launch failure or failure during other mission events. The confidence levels of known and unknown aspects of the mission concept or technology choices should be addressed. Key FOMs for the risk category include crew safety (probability of loss of crew) and mission success (probability of loss of mission). The risk models that were developed for assessing the risks include all known redundancy, reliability, and contingencies as known about the systems to date. These risk estimates will improve as the design maturity of the systems improves.

- *Hazards.* An assessment of the mission and technology risks that have the potential to cause a mishap. This includes hardware, software, and operational issues that could result in the loss of crew, personnel, vehicle, or mission. Hazard measures of effectiveness include items such as crew radiation exposure, trajectory hazards such as close passage to the sun, insertion failure during the aerocapture maneuver, etc.

- *Aborts.* An assessment of the ability of the mission concept or technology choice to provide for crew survival during various mission phases due to anomalies that result in early mission termination. Aborts could include early vehicle return or safe havens, but must result in the eventual safe return of the crew to Earth. For the most part, aborts are considered to be part of the overall risk measure, such as the ability of the crew to return to orbit due to systems failures on the surface of Mars.

- *Development.* A key FOM for the Mars architecture is development risk, which exists for the new technologies. Development risk is also associated with the design and testing of the hardware and software, beyond just the risk of successfully developing new technologies. This includes not only the flight elements, but the fabrication, test, and operations facilities that are needed to support the missions. Some factors in development risk are complexity, maturity of the technology, performance margins, manufacturability, and schedule. There are also risk factors that are not directly technical such as acquiring existing facilities, environmental approval for new facilities or modifications, planetary protection issues for Mars and Earth, potential international cooperation issues, not being able to deliver some products for the cost estimates that were committed to, and variability in the funding environment.

6.1.2 Effectiveness

Measures of performance that are associated with effectiveness focus on determining the degree to which the mission concept, or technology option, effectively meets mission needs. Future space exploration systems and missions must be effective. In other words, the capabilities of a new system or infrastructure must be worth the costs of developing, building, and owning them. The goals and objectives achieved by the missions that are using those systems and infrastructures must be worth the costs and risks that are involved in operating them. Effectiveness must be determined case-by-case, based on the specific design objectives of the system or infrastructure as well as on the detailed mission objectives (e.g., science objectives) that may be achieved.

- *Mission objectives.* Assessment of the capability of the mission approach or technology that is chosen to satisfy exploration objectives including the ability to meet scientific objectives and flexibility in mission planning and execution. This FOM includes items such as number of launches, spacing between launches, time available to support key operations, etc.

- *Mass.* Total mass that is required to be delivered to LEO to support the initial mission (includes pre-deployed infrastructure, if any) and the required mass for each subsequent mission. Also includes an assessment of the total number of launches that is required to emplace the necessary infrastructure as well as for each recurring mission. Mass measures of effectiveness also include architecture sensitivity to change in mass.

6.1.3 Affordability

To be sustainable, future space exploration systems and infrastructures as well as the missions that are pursued using them must be affordable. In other words, the costs for design, development, test, and engineering of these systems must be consistent with projected future year NASA budgets. (The same is true for the recurring costs of additional copies of all exploration systems.) Similarly, the costs that are associated with operating these systems in future space exploration missions must be consistent with projected future year NASA budgets. Assessments of affordability include the degree in which the proposed mission or technology option is expected to provide an affordable approach. Assessments in this focus area include both total expected costs as well as affordability assessments regarding expected funding profiles and phasing.

- *First mission.* Total cost for the design, development, test, and evaluation of the required systems and facilities that constitute the element or mission concept for the first human mission. This includes all necessary flights, cargo, and crew that are necessary to conduct the mission. First mission cost includes total program, infrastructure, and facility costs that are necessary for execution of the mission concept (e.g., sustaining engineering, hardware production, ground and mission operations, etc.).

- *Third mission.* Total annual program, infrastructure, recurring element, and facility costs that are necessary for execution of three complete human missions to Mars.

6.2 Decision 1: Mission Type

The choice of the overall exploration mission sequence and corresponding trajectory strategy has perhaps the greatest single influence on the resulting architecture. The ideal mission would be one that provides: (1) the shortest overall mission to reduce the associated human health and reliability risks; (2) adequate time on the surface in which to maximize the return of mission objectives and science; and (3) low mission mass, which, in turn, reduces the overall cost and mission complexity. Unfortunately the "ideal" mission does not exist, and tough choices must be made between design options. Thus, the first decision that was tackled by the MAWG addressed a key architectural component that is tied to the orbital mechanics of human Mars missions and, specifically, with the selection of the mission class, namely long surface stays vs. short surface stays. Human missions to Mars are typically classified into these two primary approaches.

Trajectories from Earth to Mars are well understood and have been used by NASA robotic mission projects for more than 4 decades. Round-trip missions to Mars and back, however, are more complex in that the outbound and inbound legs must be synchronized into an optimal mission plan. For the lower-energy outbound trajectories, upon arrival at Mars the Earth is in a relatively unfavorable alignment (phase angle) for an energy-efficient return. This unfavorable alignment results in two distinct classes of round-trip Mars missions: (1) Opposition-class missions, which are also commonly referred to as short-stay missions; and (2) Conjunction-class missions, which are commonly referred to as long-stay missions. Practical considerations, such as total propulsive requirements, mission duration, surface objectives, and human health considerations, must be considered in the mission design process when choosing between these mission classes. The period of time that is necessary for the phase angle between Earth and Mars to repeat itself varies. The mission repetition rate for identical Earth-Mars phasing and, therefore, launch opportunities for similar mission classes is on the order of every 26 months. Mission characteristics such as mission duration, trip times, and propulsive requirements vary to due to the eccentricity of Mars' orbit.

Opposition-class missions are typified by short surface stay times at Mars (typically 30 to 90 days) and relatively short total round-trip mission times (500 to 650 days). The exploration community has adopted the terminology "short-stay" missions for this class. The trajectory profile for a typical short-stay mission is shown in figure 6-2. This mission class has higher propulsive requirements than the long-stay missions, and often uses a gravity-assisted swing-by at Venus or the performance of a deep-space propulsive maneuver to reduce total mission energy and constrain Mars and Earth entry speeds. Short-stay missions always have one short transit leg, either outbound or inbound, and one long transit leg, which requires a close passage by the sun (0.7 AU [astronomical unit] or less). After arrival at Mars, rather than waiting for a near-optimum return alignment, the spacecraft initiates the return after a brief stay to make up for the "negative" alignment of the planets that exists at Mars departure. Distinguishing characteristics of the Opposition-class mission include: (1) short-stay times at Mars, (2) medium total mission duration, (3) the vast majority of the round-trip time spent in interplanetary space, (4) a perihelion passage that is inside the orbit of Venus on either the outbound or inbound legs, and (5) a large total energy (propulsion) requirement.

Conjunction-class missions are typified by long-duration surface stay times (500 days or more) and long total round-trip times (approximately 900 days). These missions represent the global minimum-energy solutions for a given launch opportunity. The trajectory profile for a typical long-stay mission is shown in figure 6-2. Unlike the short-stay mission approach, instead of departing Mars on a nonoptimal return trajectory time is spent at Mars waiting for more optimal alignment for a lower-energy return. Distinguishing characteristics of the Conjunction-class mission include: (1) long total mission durations, (2) long-stays at Mars, (3) relatively little energy change between opportunities, (4) bounding of both transfer arcs by the orbits of Earth and Mars (closest perihelion passage of 1 AU), and (5) relatively short transits to and from Mars (less than 180 to 210 days).

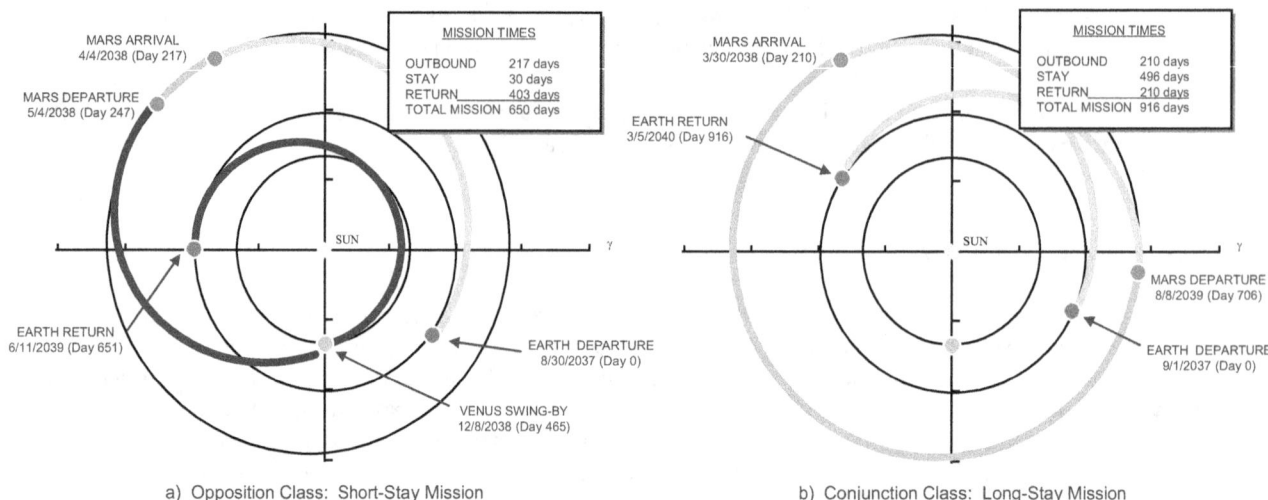

Figure 6-2. Comparison of (a) Opposition-class and (b) Conjunction-class mission profiles.

For the trajectories that were studied during 2007, the Opposition-class missions require greater total propulsive delta-V; they also experience significant variation of propulsion requirements across the synodic cycle. Variation of delta-V across the synodic cycle for Opposition-class missions is nearly 100% with an average total delta-V of 10 km/s ± 3.7 km/s. This variability significantly impacts the space vehicles, since they must be designed to provide the propellant capability and design attributes that allow for a wide range of propellant loads or the capability to deliver a wide range of payloads to Mars. There are some mission cases in which the total interplanetary delta-V is so excessive that the cases are outliers and, thus, are usually eliminated from consideration, requiring skipped mission opportunity and resulting in a minimum 26-month "stand down" before resuming the normal mission sequence. The variability of total interplanetary propulsive delta-V across the synodic cycle for conjunction class missions is fairly small, on the order of 35%, while also providing for overall lower delta-V (the average total delta-V was approximately 7 km/s ± 1 km/s). This small variation of propulsive requirement across the synodic cycle allows the use of a common vehicle and payload design for each opportunity. This common strategy also allows the vehicle systems to be flown in any opportunity, thereby reducing the potential of either skipping harder years, as in the case of Opposition-class missions, or allowing systems to be flown at a later date if necessary due to technical or schedule difficulties.

Total mission durations for the short-stay missions range from 500 to 650 days, with 30 to 90 days spent in the vicinity of Mars. For the short-stay missions, over 95% of the total mission time is spent in the deep-space zero-g interplanetary environment with the remaining 5% spent in the vicinity of Mars. The transit leg durations range from a minimum of 190 days to a maximum in excess of 400 days. The total mission durations for long-stay missions range from 890 to 950 days, with a range of corresponding surface stay times ranging from 475 to 540 days in the vicinity of Mars. For the long-stay missions, approximately 55% of the total mission duration is spent in the vicinity of Mars with the remaining 45% spent in transit. The time spent in orbit vs. the time spent on the martian surface is open to further refinement as the relative tradeoffs between mission return and crew risk are conducted.

6.2.1 Mission-class scientific position

During the deliberations on mission type, the MAWG solicited the help of the MEPAG to provide an assessment of the relative advantages and disadvantages of the two mission types under considerations. As described in Section 3, the MEPAG sponsored the creation of a special assessment group, the HEM-SAG. The HEM-SAG reviewed the proposed surface exploration strategies that are associated with both the long-stay and the short-stay mission concepts. The HEM-SAG specifically was asked to provide an assessment of the relative advantages and disadvantages of not only mission concepts that are driven by the length of stay, but also of those mission concepts that are associated with the potential return to the same exploration site or conducting subsequent missions to different exploration sites.

Short-surface-duration missions, while offering potential for breakthrough, human-enabled science are not favored for science-driven exploration for several reasons: Short-stay human surface missions could not make the best use of mobility to optimally explore a region due to the time available for EVA (and for subsurface access system operations, such as a deep drill). Short-stay human surface missions do not optimize the "iteration cycle time" that is associated with in-situ field investigations on the basis of time available (i.e., that are too few cycles in which to adapt to the unexpected scientific context that is likely to emerge). Short-stay human surface missions do not allow time for sample high-grading to ensure a best subset of materials is returned for detailed analysis on Earth. This limits the discovery potential that is intrinsic to field sampling.

Conjunction-class missions, which provide extended duration on the surface while maximizing the exploration range from the landing site, are most favored to optimize the scientific yield. A long surface stay allows maximal use of human "on-site" observational and intuitive scientific capabilities, even if EVA is restricted to 25% of the available time. By maximizing opportunities for adapting scientific investigations to a given region, the probability of paradigm-busting discoveries increases exponentially as compared to short-stay mission strategies. Long surface stay also maximizes human opportunities for using mobility (horizontal and vertical) to more completely explore a compelling region at scales that are commensurate with processes that preserve evidence of past life on Earth. In addition, the long-surface-stay scenario allows the humans who are "on site" to make best use of their non-EVA time to employ general analysis "tools" to investigate sampled materials and, hence, to best select the optimized subset (so-called splits) for the return to Earth. It should be noted that long surface stays at three independent and different human exploration sites is the most favored option.

6.2.2   Crew health and performance assessment

The Crew Health and Performance (CHP) Team of the MAWG evaluated both the short-stay and the long-stay mission architectures for the human mission to Mars. When all human health and performance disciplines were considered, no clear advantage to either option was identified on the basis of crew health, safety, and performance. A summary of the key human health and performance findings are provided in table 6-1. It is important to note that the risk assessment that was provided by the radiation discipline indicates that both the short-stay (Opposition-class) and long-stay (Conjunction-class) mission options pose a high risk that crew members would exceed current permissible radiation exposure limits, but there was a slight preference of the long-stay mission. A number of significant knowledge gaps and technologies to be developed were identified by the CHP disciplines, who concluded that no legitimate discrimination between the two scenarios would be valid, when based on that analysis with current knowledge and space flight experience, because higher-order details of the scenarios have not been fully developed. However, any Mars exploration option that is selected by NASA could be implemented concomitant with acceptance of all of the residual human health and safety risks that are identified by the CHP disciplines and their parent organizations.

6.2.3   Risk and cost assessments

Assessments of architectural crew safety (probability of loss of crew) as well as mission success (probability of loss of mission) were conducted for both the long-stay and the short-stay architectures. These comparisons must be considered first-order assessments due to the relative uncertainty resulting from the immaturity of the system concepts under consideration. End-to-end mission models were developed using "best"-known data to date, including space shuttle and ISS histories. These models were also developed "as is" with no credits taken for flight demonstrations (e.g., large-scale EDL) or other architectural activities (e.g., lunar). This process thus gives an adequate apples-to-apples comparison of the two mission classes that are under consideration.

Although the short-stay missions appear to provide slightly lower overall risk of loss of mission, there is no clear advantage given the maturity of the understanding of the systems to date. Due to the longer mission duration of the long-stay mission approach, the overall system reliability is a driver of mission success. Gaining better understanding of the system performance for long periods is necessary to reduce the risk of loss of mission. Technology and system demonstrations on the ISS and lunar programs provide a vital link to reducing this risk.

Table 6-1. Summary of Human Health Mission Type Considerations

Crew Health & Performance (CHP) Component	Short Stay (Opposition-class; 22 months total)	Long Stay (Conjunction-class, 30 months total)
Physiological Countermeasures	<ul><li>Extended 0-g transits at limits of human spaceflight experience base</li><li>Preferred option only if artificial-gravity is available</li></ul>	<ul><li>0-g transit phases well within experience base</li><li>3/8-g surface phase outside experience base, will be partially mitigated by Lunar Outpost experience</li></ul>
Human Factors & Habitability	<ul><li>Not preferred option without access to Surface Habitat</li></ul>	<ul><li>Preferred option with access to Surface Habitat</li></ul>
Radiation	<ul><li>Higher risk of carcinogenesis, acute syndromes, central nervous system effects and degenerative effects due to longer transits (solar proton events & galactic cosmic radiation) and close perihelion passage (solar proton event effects)</li><li>Option is well outside current permissible exposure limits</li></ul>	<ul><li>Slightly preferred option due to less exposure to free space heavy ion environment</li><li>Prolonged exposure to poorly-understood surface mixed-field (neutrons and charged particles) environment</li><li>Option is well outside current permissible exposure limits</li></ul>
Behavioral Health & Performance	<ul><li>Preferred option due to shorter overall duration</li><li>Possible risk due to higher acute radiation exposure within 0.7 astronomical unit</li></ul>	<ul><li>Increased risk due to longer overall duration</li></ul>
Medical Capabilities	<ul><li>Slightly preferred option due to less duration of risk exposure on surface and total mission</li></ul>	<ul><li>Slightly increased risk due to longer overall duration</li></ul>

Since the initial comparative risk models did not include flight demonstrations or the lunar program as risk mitigation steps, first use of the EDL system as well as overall system reliability are key contributors to crew safety. In addition, close perihelion passage, which is necessary for the short-stay mission approach, becomes a crew risk driver. The initial risk results indicate that the short-stay missions decrease the duration of equipment reliability, but increase the number of Ares-V launches. Certain elements are reduced with no SHAB, but cause a lack in maturity leading to greater risk for crewed missions (i.e., EDL). Equipment reliability could be enhanced by scavenging techniques when a crew is present. These techniques could be learned during lunar missions.

For the short- vs. long-stay mission, the difference in cost is due predominately to the surface systems, including the development and recurring cost of the extra SHAB, the recurring cost of an extra descent stage, the long-duration rover, the additional scientific equipment, etc. There is some uncertainty in the magnitude of the difference as some of these systems are not well-defined yet. The cost difference in the flight systems is is smaller in comparison to the cost difference in the surface systems. This is due to the modular nature of the MTVs and the similar number of total launches and flight elements. Even so, there is a slight cost savings for the short-stay flight systems and launch costs. Cost of the surface systems for the long-stay missions may be further reduced depending on commonality with lunar systems and lunar technology development activities.

6.2.4 Mission type recommendation

A summary of the overall FOMs that were considered for the long/short mission mode decision are shown in table 6-2. These results were discussed with the agency Joint Steering Group on July 23, 2007. After deliberating on the results, the Joint Steering Group concurred with the MAWG recommendation of proceeding with the long-stay (Conjunction-class) mission approach. As can be seen from this table, most of the FOMs favor the long-stay approach, with the exception of overall mission duration and a slight cost advantage. This recommendation is based entirely on our collective current understanding of system and concept performance at this time. As data are obtained and additional missions are conducted, this decision could be readdressed if warranted.

Table 6-2. Mission Type Recommendation Summary

Question	Which mission type, Conjunction class (long surface stay) or Opposition class (short surface stay) provides the best balance of cost, risk, and performance?
Recommendation	**Conjunction class (long-stay) missions**
Notable Advantages of Conjunction-Class (Long-Stay) Missions	• Best exploration value for cost. • Ample time for crew acclimation and planetary operations/contingencies and surface exploration. • Zero-g transits (~180 days) within our current experience base. Lunar outpost would provide vital hypo-gravity data for human performance associated with long surface stays for feed forward to Mars. • Less total radiation exposure (as known today – surface radiation environment characterization needed). No other significant human performance factors identified. • No close perihelion passage reduces radiation and thermal risks. • Lower total delta-V and less variation in delta-V across the synodic cycle. • Less sensitive to changes in propulsive delta-V and, thus, less architectural sensitivity. • Provides ability to maintain similar vehicle size for both crew and cargo vehicles. • Orion Earth return speed "within Orion family" – 12 km/s (TPS implications).
Notable Disadvantages	• Longer total mission duration. • Slightly higher overall total mission cost (assuming Opposition-class missions do not require dedicated surface habitats).

6.3 Decision 2: All-up vs. Pre-deploy Cargo

The issue that is associated with either pre-deploying mission cargo ahead of the crew or taking all mission cargo with the crew was the second key decision package that was addressed by the MAWG. This decision was assessed for both the short-stay and long-stay mission classes, although the discussion below focuses on the recommended Conjunction-class long-stay mission approach.

6.3.1 Pre-deploy option

For the long-stay mission sequence, two cargo elements are pre-positioned to support the crew's surface mission: the DAV and an SHAB with other surface equipment (figure 6-3). Both of these elements are launched in the same minimum energy opportunity just over 2 years prior to the launch of the crew. The launch campaign for the first two cargo elements begins approximately 8 months prior to the opening of the launch window. The cargo elements arrive at Mars approximately 8 months later and are placed in the appropriate parking orbit or at the selected surface location. They are checked for proper function and then placed into a minimal operating configuration to remain in this state for over 2 years before the arrival of the crew. The next minimum-energy window (for the next cargo elements) opens shortly before the fast-transit trajectory window for the first crew, but these launch windows are still close enough that a combined launch campaign at KSC is required. This launch campaign for the second crew's cargo and for the first crew begins as much as 1 year before either window opens so that all of these elements are ready for their respective departures. The first crew arrives before the cargo elements for the second mission and nominally uses the assets that were launched over 2 years previously. However, should either the DAV or the SHAB suffer a failure between the time the first crew launches from Earth and the time at which they leave Mars to return to Earth, the second set of cargo elements could be used, thus potentially preventing loss of the mission or of the crew. This is a unique feature of the pre-deployment strategy when applied to the long-stay mission; this overlap of assets is not available for any of the short-stay options or for the all-up strategy.

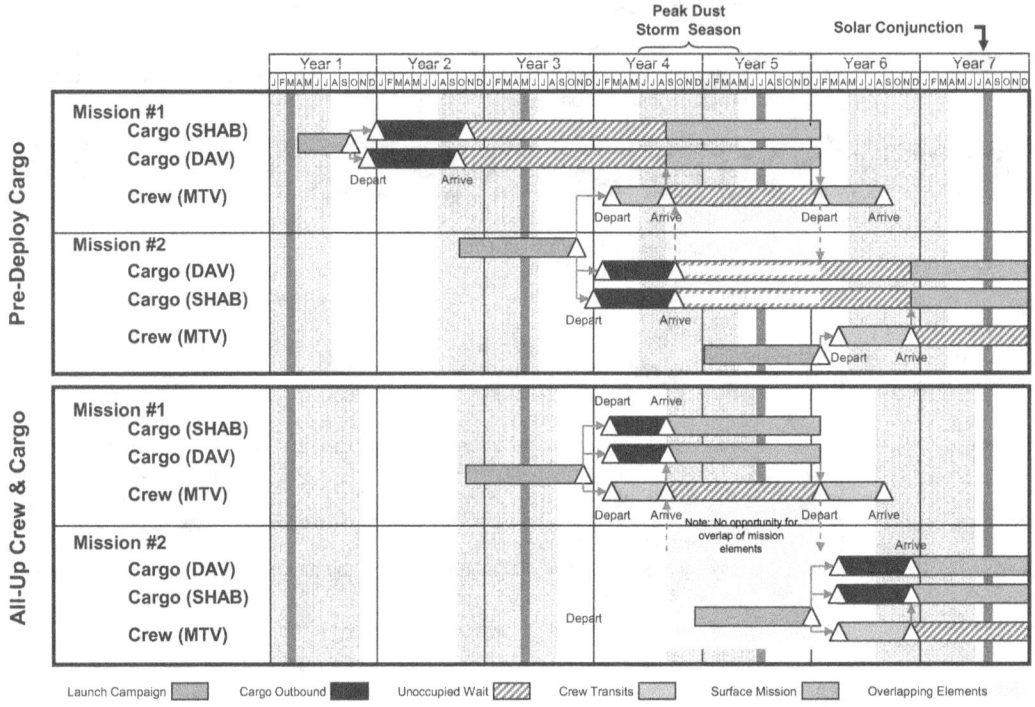

Figure 6-3. Example mission timeline comparison for Conjunction-class missions.

6.3.2 All-up mission option

For the long-stay mission sequence, two cargo elements are required to support the crew's surface mission: the DAV and an SHAB with other surface equipment. All of these elements are launched on a fast transit trajectory so that they all arrive at Mars at the same time. While it is conceivable that all of these elements could be integrated into a single stack while in LEO, the total mass of such a stack would be quite significant (i.e., equivalent to several ISSs) and likely difficult to control. The total thrust that is required to avoid significant gravity losses during departure also makes this approach less desirable. The alternative – three closely spaced departures from LEO during the same launch window followed by a rendezvous (but not necessarily docking) in interplanetary space – is also not trivial but is considered manageable and, thus, would be the preferred approach for this option. The KSC launch campaign begins approximately 1 year before these elements depart for Mars; this is similar to the situation described for the pre-deploy strategy. The launch campaign for the next mission begins approximately 1 year after completion of the first campaign. There is no overlap at Mars of the two crews or their equipment.

6.3.3 Cargo mission mode recommendation

Based on the analyses and results presented here, the study team recommended that the pre-deploy feature be used as the reference approach in DRA 5.0. These results were discussed with the agency Joint Steering Group on July 23, 2007. After deliberating the results, the Joint Steering Group concurred with the MAWG recommendation of pre-deploying mission cargo to Mars one opportunity before the crew. An overview of the key advantages and disadvantages is provided in table 6-3.

Table 6-3. Cargo Deployment Recommendation Summary

Question	Should mission assets, which are not used by the crew until arrival at Mars, be pre-deployed ahead of the crew?
Recommendation	**Pre-deploy cargo one opportunity (26 months) ahead of the crew**
Other Questions	• Is a lifeboat mode (e.g., Apollo 13) feasible/advantageous for human Mars missions? • What are the architectural advantages of all-up vs. pre-deploy mission modes?
Notable Advantages of Pre-deployment	• Enables strategies such as ISRU. • Mission design provides natural functional redundancy to reduce crew risk. • Verifies cargo arrival at Mars and operational condition prior to crew departure from Earth. • Satisfies more exploration goals via robotic exploration prior to crew arrival. • Provides lower total initial mass in LEO. • Reduces outbound vehicle size and complexity.
Notable Disadvantages	• Longer cumulative time on systems. • Slightly higher costs (mission operations time).

6.4 Decision 3: Aerocapture vs. Propulsive Mars Orbit Capture of Cargo

To place a spacecraft in orbit around a planetary body, sufficient velocity must be removed such that the gravitational field of the target body would transform the approach hyperbolic trajectory into a closed elliptical orbit. Traditionally this has been accomplished using chemical propulsion to provide deceleration forces to slow the spacecraft to the required velocity for orbit capture. For planetary bodies that possess an atmosphere, including Mars, using atmospheric drag to provide aerodynamic deceleration ("aerocapture") may result in significant mass savings as compared to the more traditional propulsive orbit insertion method. The use of aerocapture for the Mars mission cargo elements was the third decision package that was assessed by the MAWG.

6.4.1 Overview of aerocapture

Aerocapture is a method that is employed to directly capture into a planet's orbit from a hyperbolic arrival trajectory using a single, atmospheric aerodynamic drag pass, thereby reducing the propellant required for orbit insertion. Over the last several decades, multiple aerocapture systems analysis studies have been conducted for multiple planetary destinations (Earth, Mars, Venus, Titan, and Neptune) that have used a variety of aerodynamic shapes and guidance algorithms; all have concluded that aerocapture is a moderate- to relatively low-risk technology. However, these studies were typically limited to the significantly smaller payloads (1 to 2 t) that are associated with robotic missions. This effort attempted to address aerocapture performance for the much larger 50- to 100-t payloads that are required for human-class missions. The aerocapture technique requires an aeroshell with sufficient TPS to protect the payload from the aerodynamic heating that is encountered during the atmospheric pass. During the aeropass maneuver, an atmospheric flight guidance and control algorithm is used to target the trajectory to a specified condition following atmospheric exit; then an orbit periapsis raise maneuver is executed to achieve the target orbit conditions, as shown in the aerocapture flight profile schematic in figure 6-4.

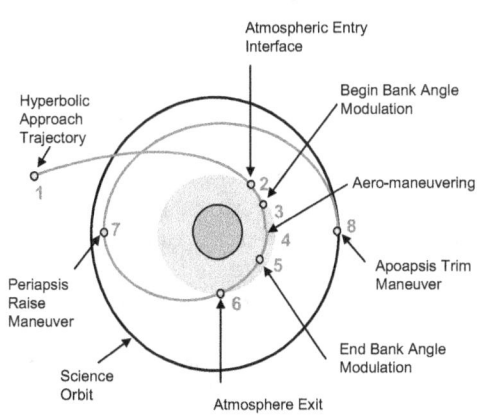

Figure 6-4. Aerocapture flight profile.

Although the aerocapture technique has not yet been demonstrated on an operational mission, studies have demonstrated its feasibility and identified potential savings in the propellant and overall system mass required for orbit insertion. The primary aerocapture technology challenges are the TPS, sufficient knowledge of the atmospheric density profiles, and aerocapture guidance and control algorithms. TPS challenges are thought to be no more demanding than direct entry TPS, but are configuration-specific to new shapes and heat-pulse duration.

Given that the Mars EDL system would already require a hypersonic entry aeroshell for the landing portion of the mission, it could be easily modified to also serve as the aerocapture aeroshell by adding the appropriate TPS to that which already exists for the hypersonic entry phase of the mission. In fact, this and a small amount of additional propellant for the post-aerocapture periapsis raise and orbit trim burn are the only major hardware additions to the system to enable an aerocapture maneuver. The question then is: What are these additional mass requirements, relative to the propellant mass requirements, for the propulsive orbit capture?

6.4.2 System-level trades

To determine the potential mass savings, system-level trades were conducted using aerocapture and both chemical and NTP options for MOIs. First, it was important to understand the sensitivities of aerocapture performance to the possible variations in vehicle design for this mission. The key parameters of interest were ballistic number, L/D, target orbit (500-km circular orbit, or a 1 Mars sol orbit period), and the atmospheric entry velocity of the arriving vehicle. Early in the study, the specific EDL system definitions were not yet fully defined; so to investigate the scope of the problem, several initial conservative assumptions were made. The desired useful landed payload mass at the surface of Mars was assumed to be between 20 t and 80 t. These values were used to derive an entry mass using a set range of "gear ratios" that was obtained from historical sources. Based on the assumption of no EVA on-orbit assembly of components, aeroshell dimensions were assumed to be constrained to the estimated capability of the launch vehicle. The reference Ares V launch vehicle payload shroud provided accommodation for a 7.5-m-diameter and 12-m-long vehicle. Potential growth was estimated to 12 m in diameter and 35 m in length. To limit the scope of the study, initial geometric assumptions were based on an ellipsled configuration for the aeroshell; however, given the same dimensions, small modifications to the trim angle-of-attack could be made to achieve similar ballistic numbers and L/Ds for other shapes, including the biconics and triconic class of slender body mid-L/D designs.

To understand the benefits and consequences of performing aerocapture at various L/Ds and ballistic numbers, a parametric aerocapture assessment assuming optimal performance was employed. This analysis indicated that the open-loop guidance assumption was valid, with a noted variability in minimum altitude of approximately ±2 to 3 km. In addition, the analysis indicated that there is an adequate entry flight path angle margin (skip out margin on the order of 0.10° 3-σ,) for the entire range of possible vehicles as estimated navigation errors. Note that for previously flown robotic missions, the Mars approach margin is approximately 0.25° 3-σ. The peak heat rate and total integrated heat load sensitivities were also assessed as a function of the aeroshell L/D and ballistic number for the aerocapture maneuver. These data indicate the fact that for moderate ballistic numbers (400 to 1,000 kg/m^2), the peak heating and total heat loads are well bounded by the TPS performance capabilities that are being developed for the Orion CEV lunar return conditions.

6.4.3 Aerocapture for the reference payload

An aerocapture point design solution was developed for the reference 40-t useful payload on the surface of Mars. Corresponding to this was an EDL system design that resulted in a Mars arrival mass of approximately 115 t. Monte Carlo simulations were used to conduct capability verification. The results indicate that there is adequate targeting performance for this vehicle despite the high ballistic number (498 kg/m^2) and the high-energy orbit (exit velocity representing 97% of escape velocity), which increases the difficulty of targeting the desired orbit. Although the standard deviation on the apoapsis dispersion is approximately 1,400 km, further tuning of the guidance algorithm would improve this performance. The overall performance of the aerocapture maneuver can be measured in terms of the post-aerocapture circularization burn requirements, which indicates that the mean delta-V that is required is only 19 m/s, with a maximum 66 m/s case. Based on the results of these parametric and Monte Carlo performance assessments, the aerocapture maneuver was deemed to be a feasible option for the large-scale, high-mass systems that are consistent with the human-class mission set.

6.4.4 Aerocapture vs. propulsive capture comparison

Both propulsive and aerocapture Mars orbit insertion were examined and compared for both the NTR and conventional chemical propulsion approaches. The aerocapture cases required slightly more mass in Mars orbit (4 t) due in large part to the additional TPS that is required to execute the aerocapture maneuver, which would then be reused during the EDL phase. Detailed TPS sizing assessments were performed for both the aerocapture followed by entry and the entry alone options to validate these data. However, the masses were small in comparison to the additional NTR and chemical propellant masses that are required to execute the all-propulsive burns for orbit insertion. Ultimately, the performance metric that was used for direct comparison was determined to be the IMLEO. The missions that use aerocapture achieve a significant savings in IMLEO requirements. The analysis indicated a mass savings of 350 t when using aerocapture as compared to chemical MOI, and a savings of approximately 87 t for aerocapture vs. propulsive capture for the NTR case.

Risk and cost assessments were also conducted for the aerocapture vs. propulsive MOI options. Detailed quantitative loss of mission and loss of crew risk metrics were difficult to define; however, qualitative assessments of the risks that are associated with aerocapture were reviewed. Many systems analysis studies and projects have examined multiple targets (Earth, Mars, Venus, Titan, Neptune), with a variety of shapes (low L/D sphere cones to mid-L/D slender bodies), with/without aerodynamic control surfaces, and a variety of guidance algorithms, and have all concluded that aerocapture is a relatively low-risk technology. The overall TPS requirements for Mars aerocapture are much less stressing than those associated with either lunar or Mars Earth return. Given the similarities between the aerocapture MOI maneuver and the skip entry maneuver that may be used by the Orion CEV for lunar return, many of the risks that are associated with aerocapture, including guidance system performance and dual-use TPS, will be retired via the CEV/Orion development program. The use of aerocapture is felt to be a relatively small incremental cost to the larger, more challenging EDL system development costs and risks. The major engineering challenges and technology risk reduction efforts that are required for EDL system development will also serve to retire many of the risks that are associated with aerocapture technology. Some incremental technology development and risk reduction efforts will be required, but these are felt to be moderate and easily manageable. Preliminary risk analysis and modeling indicate that there are no significant risk discriminators between aerocapture and propulsive capture (chemical or NTR).

The cost assessments that were performed indicate that there is a distinct long-term cost advantage to the aerocapture mission for the chemical propulsion option due to a large reduction in the number of launches and flight elements. The cost for the multi-use aeroshell design and production for three missions was estimated as a 6% increase over the cost of an entry-only aeroshell. There is some cost risk inherent to this assessment due to the technological uncertainties that exist in the development of the dual-use aerocapture aeroshell, but this was deemed to be small and the overall cost sensitivity to this assumption was minimal. The cost advantage of the aerocapture option is reduced for the NTR-based propulsion due to the reduction in mass sensitivity that NTR provides. For the NTR systems, cost is not seen as a significant factor in the aerocapture trade. Aerocapture does provide increased launch margins that may have cost implications that are not captured.

6.4.5 Aerocapture recommendation

Based on the analyses and results that are presented here, the study team recommended that the option to aerocapture Mars cargo elements into Mars orbit be retained. These results were initially discussed with the agency Joint Steering Group on July 23, 2007, and then again in their final form on January 17, 2008. After deliberating on the results, the Joint Steering Group concurred with the MAWG recommendation of aerocapture of cargo elements into Mars orbit. An overview of the key advantages and disadvantages is provided in table 6-4.

Table 6-4. Aerocapture Recommendation Summary

Question	Should the atmosphere of Mars be used to capture mission assets into orbit (aerocapture)?
Recommendation	**Retain aerocapture for Mars cargo elements**
Notes	• Benefit of aerocapture is dependent on the interplanetary propulsion that is used. (If NTR is used, the issue becomes one of risk. If chemical is used, aerocapture was considered enabling.) • Aerocapture for the crew transfer vehicle was eliminated from consideration due to the physical size of that element.
Notable Advantages of using Aerocapture for Mars Orbit Insertion	• Aerocapture reduces total architecture mass. • Less architecture sensitivity to changes in payload mass. • Minimal TPS impacts. Both heat rate (factor of 3) and heat load (factor of 2) are less than those that will be experienced for the Orion Earth return mission. • Aerocapture guidance techniques are subsets of Orion skip trajectories.
Notable Disadvantages	• Dual use of TPS (aerocapture followed by EDL) increases overall risk. • Heat rejection and thermal load on primary structure are yet to be assessed and will add mass and complexity.

The reference aerocapture system architecture is felt to be a conservative design that includes a large, nonoptimized aeroshell with relatively large uncertainty margins, including TPS. However, mass estimates are based on engineering judgment and extrapolations primarily from much smaller-scale robotic EDL systems. There are other options yet to be explored, including dual-use launch vehicle/aerocapture shrouds, inflatable/deployable aeroshells, etc., that may substantially improve the performance of the aerocapture and EDL system as a whole. Therefore, the recommendations for the development and use of aerocapture technology for the human Mars architecture are as follows:

1. Continue to include aerocapture for MOI on cargo missions as the reference approach until a decision is made on the propulsion option (chemical vs. NTR).

2. Conduct detailed "pre-phase A" point designs to validate mass models for both aerocapture and propulsive capture MOI.

3. Continue to pursue options to improve aerocapture system performance and understand overall system-level performance, cost, and risk sensitivities and drivers.

4. Take advantage of the Orion/CEV lunar return "skip entry" qualification and flight data to retire risks that are associated with dual-use TPS and aerocapture guidance performance.

It must be noted that time constraints of the study limited detailed assessments of integrated systems design impacts including thermal soak-back, center of gravity control, and separation dynamics, to name a few. Further assessments in these areas are necessary to adequately address the use of aerocapture techniques for capturing cargo elements into Mars orbit.

6.5 *Decision 4: In-Situ Resource Utilization for Mars Ascent*

Mars ISRU involves the production of critical mission consumables, such as propellant and life support consumables, from resources that are available at the site of exploration. The main rationale for incorporating ISRU technologies into a Mars mission is to attempt to reduce the IMLEO by reducing landed mass (IMLEO being a first-order measure of cost and risk). Incorporation of ISRU could also significantly enhance, if not enable, more robust exploration capabilities while also providing redundancy of critical functions such as life support. Since propellants and life support consumables for a long surface stay make up a significant fraction of the mass that must be launched from Earth, ISRU could either reduce the total amount of mass that must be launched or replace propellant and consumable mass with extra payload or science. In particular, the potential benefits of ISRU were assessed for the Mars ascent vehicle propulsion system and for the creation of consumables for life support and EVA needs. Several ISRU technologies were analyzed for their mass-reduction benefits during the course of trade studies for DRA 5.0. Analyses must take into account the mass of all hardware that is needed to enable ISRU (including power systems), the total volume (including reagents brought from Earth), and any risk the use of ISRU contributes to loss of mission or crew. Past DRMs have also documented some of the benefits of ISRU technologies, but in a less comprehensive manner than is documented here. Prior studies were limited to the investigation of Mars atmospheric resources (e.g., CO_2, N_2, and Ar), whereas this study also performed an initial investigation of the use of surface regolith material as a source of H_2O as well. Mars ascent vehicle propellant options of LOX/CH_4, LOX/H_2), and hypergolic propellants have all been examined and traded in the past.

6.5.1 In-situ resource utilization operational concept

It is important to note that the use of ISRU for ascent propellant necessitates a different operational concept than sending a fully fueled ascent vehicle to Mars. The key salient differences focus on which vehicle is pre-deployed and which vehicle the crew lands in; namely:

- *No ISRU*
  - SHAB pre-deployed to the desired landing site on Mars.
  - Crew lands (with ascent propellant) near the surface habitat.
- *With ISRU*
  - Ascent vehicle (without propellant) is pre-deployed to the desired landing site on Mars. Propellants are produced prior to the crew leaving Earth.
  - Crew lands in the SHAB near the ascent vehicle.

One often perceived drawback of the ISRU propellant strategy is the lack of abort-to-orbit (ATO) capabilities that are inherent in the ISRU propellant-derived vehicle. The key leverage of the in-situ propellant production strategy is derived from the fact that ascent propellants are made at the planet (in-situ), thus dramatically reducing the overall transportation mass that is required. This results in a lander vehicle that could not perform ATO maneuvers during the landing sequence. The ATO strategy has been a risk-reduction philosophy that has been followed since the early days of human exploration. During critical mission maneuvers, abort strategies with well-defined gates and sequences are established such that, if warranted, they could be exercised to place the crew in a stable position, namely in orbit. With the Mars in-situ propellant production strategy, ATO scenarios do not exist since the ascent propellants are produced on the surface of Mars and are not transported with the crew.

This lack of ATO capability that is inherent with in-situ propellant production has led many to discount the overall strategy of ISRU. During development of the DRA 5.0, the specific question of ATO was raised. The EDL community reviewed the typical entry sequence and concluded that, due to the physics involved during the atmospheric entry phase, ATO was probably not possible; and if it were required, it would only be available during the final portion of the entry sequence, namely the terminal phase after separation from the aeroshell had occurred near the surface. At that point, the most critical phases of the entry maneuver have been completed. Thus, emphasis of the EDL philosophy changed from one of ATO, to an *abort-to-surface strategy*; that is, to provide enough functionality and reliability in the EDL system to enable a safe landing on the surface and subsequent rendezvous with the ascent vehicle. In this sense, the final landing accuracy must be within a distance that is accessible by the crew, which includes the distance that a rover, taken with the crew, could reach.

6.5.2 In-situ resource utilization trades performed

ISRU propellant production has two main influences on mission architectures; these are to reduce (1) the mass and volume of the lander, and (2) the propulsive needs for both Mars capture and Mars departure by enabling higher rendezvous orbits compared to non-ISRU missions. Previous mission evaluations covered the first main influence but ignored the second main influence. Also, since the last human Mars mission study was performed, NASA and ESA orbital and surface robotic missions have determined that H_2O is globally available in the Mars soil in varying concentrations and depths. Therefore for the first time, Mars H_2O was considered as a potential resource in this study with and without the use of Mars atmosphere resources. For DRA 5.0, several trade options for ISRU and their impact on mission mass/volume and optimum technology/ISRU process were evaluated to understand all of the potential mission implications and benefits of incorporating ISRU into human Mars exploration plans (figure 6-5). ISRU process systems and subsystems were developed for each of the trade tree branches. For atmospheric processing options, previous models were updated with new technology and hardware performance information. For Mars soil/water processing, new models for excavation and soil processing were generated from recently created lunar ISRU regolith and processing models with Mars soil/water parameters applied. While not perfect, this allowed for first-order evaluation of the system mass, volume, and power that are associated with Mars water resource collection.

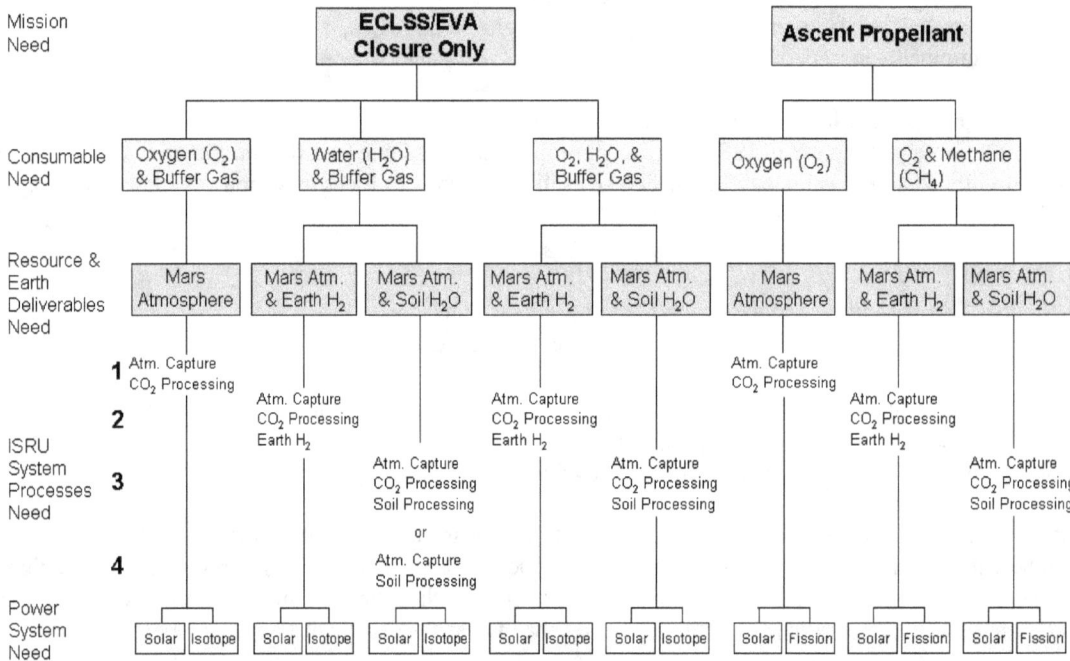

Figure 6-5. ISRU trade tree.

6.5.3 In-situ resource utilization trade study results

The mass, power, and volume calculations for the three main ISRU options for meeting all mission consumable needs are depicted in figure 6-6. These estimates assume that nuclear power is available for continuous operation. Production of O_2 alone is the highest mass system (because it requires delivery of CH_4 fuel from Earth), but it also has the lowest volume and power requirement. Production of O_2 and CH_4 from the martian atmosphere (with H_2 from Earth) results in a lower total mass than O_2 production alone; however, the volume is significantly higher than all other options. Production of both O_2 and CH_4 with atmosphere and soil/water resources is the lowest mass but highest power option. It should be noted that by changing the soil H_2O content assumption from 3% to 8%, both volume and power that are associated with the combined atmosphere and soil resource ISRU option were significantly reduced. Although the soil-based approach looks promising from a total mass and volume perspective, numerous significant challenges that are associated with operation of the required equipment, such as excavators and haulers, remain.

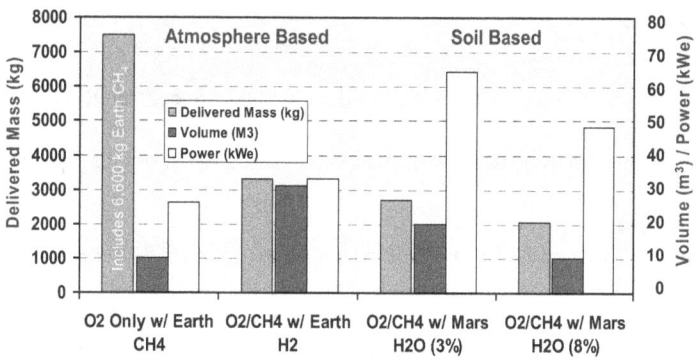

Figure 6-6. Mass, power, and volume of ISRU strategies.

6.5.4 In-situ resource utilization recommendation

The ISRU trade study results were discussed with the agency Joint Steering Group on January 17, 2008. After deliberations on the results, the Joint Steering Group concurred with the MAWG recommendation of incorporating atmospheric-based ISRU (O_2) for Mars ascent as well as consumables (O_2, H_2O, and buffer gases) for the surface mission. Using O_2 for Mars ascent provided the best balance between mass savings, total volume required, and power generation needs. An overview of the key considerations is provided in table 6-5. Fewer launches have the greatest effect on bringing down both risk and cost. Since incorporation of ISRU results in lower IMLEO and resulting lower architectural gear ratio, this leverage could be used to reduce the performance requirements on the landers (e.g., accepting lower specific impulse in favor of higher reliability engines). For both ascent propellant and ECLSS consumables, the use of ISRU necessitates greater power, which comes at the cost of mass. This is one apparent disadvantage of using ISRU. However, significant surface power capability would be required to support the crew activities and surface infrastructure regardless. Likewise, using the more traditional approach of landing a fully fueled Mars ascent vehicle would make surface operations (e.g., transfer of consumables) simpler and, therefore, more mass efficient. Another important FOM is a catch-all termed "mission flexibility." ISRU gives the ability to produce fuel for roving, EVAs, and other activities that would otherwise be limited by a fixed consumables budget. This additional flexibility provides an advantage over the non-ISRU case. Further assessments in the areas of defining the global distribution and concentration of H_2O in the form of hydrated minerals, along with concepts for excavation and soil/water processing on Mars, as well as evaluation of the potential advantages of pressure-fed propulsion options using all propellants that are produced from martian resources should be pursued in the future.

Table 6-5. ISRU Recommendation Summary

Question	Should locally produced propellants be used for Mars ascent?
Recommendation	ISRU (production of O_2 from the atmosphere) for ascent from Mars as well as consumables for the crew is enabling for robust human exploration missions
Notable Advantages of ISRU	• Production of O_2 from the atmosphere for ascent from Mars as well as consumables (O_2, buffer gases, H_2O) for the crew enables robust exploration. • Atmospheric-based ISRU processes are less operationally complex than surface-based processes. • Reduced total initial mass in LEO and subsequent number of launches. • Reduced lander vehicle size and volume. • Greater surface exploration capability (EVA, roving, etc.). • Life support functional redundancy via dissimilar means. • Lower mission risk due to fewer launches. • Lower life cycle cost through third mission (if same landing site).
Notable Disadvantages	• Requires slightly more peak power. • Longer cumulative time on systems. • Rendezvous with surface ascent vehicle required for crew return to orbit (see note).
Note	• ATO during EDL deemed not feasible. Thus, for human exploration of Mars, emphasis should be placed on abort to surface and landing accuracy.

6.6 Decision 5: Mars Surface Power

The last major decision that was addressed by the MAWG during 2007 concentrated on surface power strategy. A major consideration in developing the power trade is the power requirement that is imposed by the mission. For the mission architectures that were considered, two major phases are defined, each with separate power requirements.

6.6.1 In-situ resource utilization phase

The first phase, which commences shortly after the landing of the cargo vehicle and extends to the arrival of the crew, is the ISRU phase. During this period, power must be provided to process in-situ resources either for crew and EVA consumables only (O_2 and N_2/Ar) or for propellant O_2 production in addition to consumables. Estimates for power requirements for these two scenarios vary with the assumed power source. For the case of nuclear fission power, it is assumed that the ISRU plant would be operated continuously for at least 300 days to produce the necessary resources. In the case of solar power, the total energy would be the same but operation of the ISRU plant would be limited to 8 hr/day, at three times the power level of the nuclear case. This daytime-only operation avoids the need for the large quantities of fuel cell reactants that would be necessary to provide round-the-clock production; but, in turn, substantially larger surface arrays must be packaged, outfitted on the cargo lander, and then deployed. Daytime-only production may also result in inefficiencies that will need to be evaluated further to determine whether additional margin should be provided to the solar power requirement. Current estimates place the power requirement for the consumables-only ISRU case at 2 kWe continuous, or 6 kWe for 8 hr/day operations. When O_2 propellant production is added, these power requirements rise to 26 kWe and 96 kWe, respectively.

6.6.2 Crewed phase

The second major phase of the mission is the crewed phase, which commences with arrival of the crew at the outpost site. Power requirements for this phase vary among the three scenarios that were considered for the mission architecture, depending on extent of mobility provided and the presence of a dedicated habitat. After completion of the ISRU consumable production, the total power requirements drop, with total continuous power loads during the crew phase requiring about 17 to 20 kWe.

6.6.3 Solar power system concept

Although solar arrays face a number of challenges on Mars, the relative simplicity and technical maturity of PV systems makes them a candidate for application even to large-scale human missions. For the present study, it was decided that an optimum approach for solar power would be to develop a modular PV system that would be capable of providing 5 kWe continuously. An optimal number of these units could be deployed to provide the power that is necessary to support base operations. Additional units could be provided for redundancy. An artist's illustration of a power system consisting of five of these 5 kWe modules is shown in figure 6-7. Each module would consist of one or more solar array wings providing a total of 290 m^2 of solar array area. Solar arrays would be populated with 29% efficiency triple junction cells. The arrays would be fixed at an inclination angle that would allow evening out-of-power output over the course of the day, and would facilitate automated dust removal systems. The solar arrays feed power to a central box that contains power management and distribution equipment, as well as five RFCs that would provide 5 kWe of power for nighttime operations. These five modules would also be sufficient for consumables-only or propellant production ISRU, operating at 100 kWe for 8 hr/day while supplying 3 kWe nighttime power through the RFCs, even if a dust storm should cause ISRU operations to be suspended for up to 50 days. The mass of a single 5-kWe module is estimated to be 2,919 kg (including 20% contingency). In addition to the nominal power needs, additional power generation must be supplied in the event of a dust storm during the crew phase of the mission. For this situation, it is envisioned that the crew would bring with them thin film arrays that would be deployed by the crew on arrival. The mass and total area for this system was estimated at 7,800 kg and 4,300 m^2, respectively. The addition of emergency power generation would bring the overall solar power system mass to about 22,500 kg (including 20% contingency).

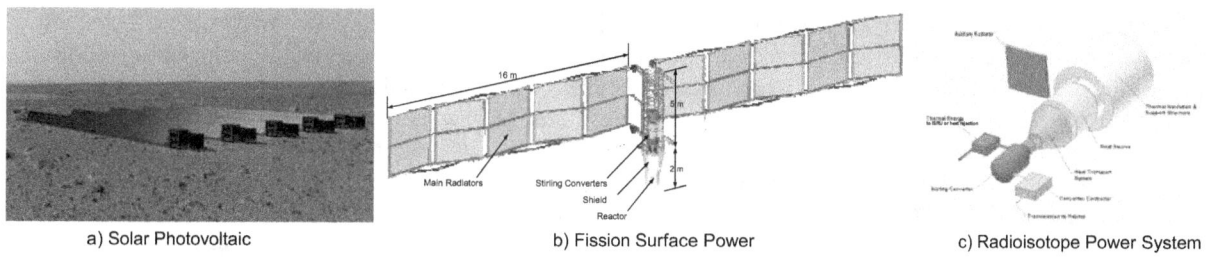

a) Solar Photovoltaic   b) Fission Surface Power   c) Radioisotope Power System

Figure 6-7. Surface power concepts.

6.6.4 Fission Surface Power System

The Fission Surface Power System (FSPS) design is taken directly from recent work performed to develop a low-cost, low-temperature system for the lunar architecture. One of the key features of this power system design is that it is adaptable to use either on the lunar or the martian surface. A sketch of the 40-kWe design that was developed for the moon is shown in figure 6-7. The reactor itself is located at the base of the power system, situated below grade. For the Mars case, an above-grade option was selected that is surrounded by a radiation shield, preferentially thicker in the direction of the base. This design allows the reactor to be sited at a 1-km separation distance from the base, achieving a dose rate during reactor operation of <5 rem/yr at the vicinity of the base. The shaped shield provides a dose rate of <50 rem/yr in all other directions. Implementation of the FSPS would require the reactor to be landed with its own mobility system, or use a separate power cart (preferred option) that would autonomously drive the FSPS to a distance of 1 km from the landing site, deploying a power cable as it traverses. Once the implementation site is reached, the FSPS would deploy its radiators and start up of the reactor would be performed. From the end of start-up operations, full power would be available to the base essentially independent of time of day or atmospheric conditions. The estimated mass for a 20-kWe reactor that might be used for the non-propellant ISRU cases is 6,800 kg (including 20% contingency). The mass for a 30-kWe reactor that could accommodate propellant ISRU is estimated at approximately 7,800 kg.

6.6.5 Large-scale Radioisotope Power System

An additional power system concept that could be considered for applications such as backup power and mobility is the large-scale RPS. Large-scale RPS designs, which are based on Stirling engine technology, have been under development in power levels up to 10 kWe. For this study, a 5-kWe RPS has been considered as shown in figure 6-7. This system consists of a heat source that is made up of 54 general-purpose heat source (GPHS) modules containing a total of 32.4 kg of plutonium (^{238}Pu). For comparison, this is the same amount of ^{238}Pu that is currently being used to power the Cassini spacecraft at Saturn. While the Cassini radioisotope thermoelectric generators (RTGs) are able to provide 1 kWe to the spacecraft, the much greater efficiency of the Stirling generator would enable such a large-scale RPS to generate 5 kWe from the same amount of fuel. The RPS would provide a continuous power source from the time that it is fueled, with a power output that is estimated to fall off by about 0.8% per year as a result of natural decay of the ^{238}Pu fuel. The current design for the 5 kWe RPS has an estimated mass of about 450 kg (including 20% contingency). A smaller 2.5 kWe system has a mass that is estimated to be approximately 230 kg.

6.6.6 Implementation considerations

Accumulation of dust on both horizontal and vertical surfaces has been a salient feature of Mars surface missions. The MER and the earlier Mars Pathfinder mission witnessed power output drops of 0.2% per day resulting from dust. The surprising longevity of the MER has been a result of "clearing events" seen by both rovers that temporarily mitigated the output losses; however, dust buildup has been seen to resume following these cleaning events. The design of a solar power system that is critical to mission success would not be able to rely on these cleaning events, which are incompletely understood. It would be necessary for a solar-powered system to incorporate some form of autonomous dust mitigation technology prior to crew arrival. Dust deposition would have a minimal effect on the FSPS. Dust adhesion to the radiator surfaces could potentially result in a slight decrease in emissivity, but this should not significantly affect operation.

Perhaps the greatest threat to the solar-powered system is the incidence of large-scale dust storms on Mars. Regional and global dust storms could dramatically reduce the amount of sunlight reaching the surface, reducing solar power to a fraction of its nominal levels. Recent experience on the MERs has shown a decrease in the power output during the worst days of the storm, down to 15% of pre-storm capability. The solar-powered system must be designed to provide at least minimal survival power during dust storms, which may last for 1 to 2 months. This could be provided by including extra solar array area, additional fuel cell capacity, or a combination of both. Analyses for the current study included sizing for crew survival during a dust storm with a duration of as many as 50 days. This would require an additional solar array of approximately 4,300 m^2 to be deployed prior to the dust storm. The concept that was considered for the study would entail having the crew deploy a thin-film array blanket to provide extra power for the duration of the storm. (However, if a second storm were to occur, the array might not be capable of producing sufficient power). Mass of the extra blanket is estimated at about 7,800 kg. Again, the dust storm conditions would have little effect on the FSPS. The radiator would see the daytime temperatures drop while nighttime temperatures would increase, resulting in no significant change in overall performance.

A major distinguishing feature of the FSPS is its ability to operate at any latitude on the martian surface. The solar-powered system, however, would be more limited in its geographical range. Previous studies have shown that the applicability of a solar-powered system is best between latitudes of about 15°S and 30°N, with system effectiveness falling off quickly beyond this region. Of the 58 sites of potential interest that were identified by the HEM-SAG, approximately 26 to 28 fell within the latitudes where solar power would be a viable option.

6.6.7 Surface power system recommendation

Based on the power system evaluations and the strengths and weaknesses that are associated with the various power concepts, including the incorporation of ISRU in the reference architecture, five decisions were made with respect to the surface power recommendations.

These results were discussed with the agency Joint Steering Group on Jan. 17, 2008. After deliberating on the results, the Joint Steering Group concurred with the MAWG recommendation of incorporating nuclear surface power for the surface mission. An overview of the key considerations is provided in table 6-6.

Table 6-6. Power Recommendation Summary

Question	Which surface power strategy provides the best balance of cost, risk, and performance?
Recommendation	The FSPS is enabling for the human exploration of Mars
Notable Advantages of Nuclear Surface Power	• Enables ISRU strategies. • Reduces power system mass and corresponding total mission mass. • Less sensitive to increase in power loads. • Continuous high-power generation. • Low sensitivity to environmental effects such as dust storms. • No restrictions to landing site location. • Less complex autonomous system deployment. • Has potential for synergism with lunar power approach and testing to reduce risk. • Lower overall cost (assuming lunar development).
Notable Disadvantages	• Inability to repair power-generation system. • Increased crew radiation dose as well as operational keep-out zones. • Increased development and testing complexity.

1. Any power architecture that is implemented upon which the crew depends should incorporate a reliable backup power system that is capable of supplying survival power. This could be accomplished by using two or more FSPSs (the DRA 5.0 strategy), through the incorporation of one or more large-scale RPSs (DIPS), or it could be provided by a sufficiently robust auxiliary system that has adequate energy storage capacity.

2. It is recommended that the FSPS be the primary power source for mission scenarios that incorporate propellant ISRU. It is recognized that while these missions could also be implemented using solar power, subject to latitude restrictions, the FSPS would result in a significantly lower mass and an arguably simpler implementation, providing a steadier and more robust power source that would benefit the ISRU process efficiency. Additionally it is felt that the complexity and criticality of successful deployment of the full complement of solar arrays that is needed for the ISRU phase would pose a significant challenge.

3. Reliability and cost estimates for nuclear power are strongly dependent on the development and test programs that precede the Mars mission. Development of a nuclear system and use in the lunar environment would greatly reduce the cost and significantly reduce the residual uncertainties in long-term operational reliability that would remain subsequent to terrestrial developmental testing.

4. The power estimates for crew operations in this study did not take into account the possibility of nighttime or emergency power modes that could be implemented in case of a dust storm. Such a low power mode could greatly reduce the additional solar array area that is needed to accommodate dust storm periods and, thus, simplify the solar option.

5. It is noted that while the FSPS design that was considered in this study was the product of a fairly detailed design study that was performed for the LAT, the solar power system did not benefit from such an effort. Any further consideration of solar power systems should begin with a design study that would develop a detailed implementation concept that is more fully tailored to this application.

7 KEY CHALLENGES

One of the principal challenges of future human exploration of Mars is to build a program that is credible in costs and schedules for critical near-term technology development and, at the same time, that addresses the broader spectrum of technologies that enables longer-term program goals. Success in meeting this challenge depends upon a solid understanding of state-of-the-art engineering systems, as well as a feasible projection of what could confidently be achieved through focused research, technology development, and system-specific advanced development programs.

Mobile Home – An artist's concept depicting long-range exploration by using pressurized rovers on the surface of Mars. Rawlings 2007

Human exploration of Mars would build directly on the technologies that are developed for and proven during the robotic phases of the Mars robotic program and lunar outpost missions. Sending humans to Mars would mandate developments in almost all areas of technology as well as in understanding system performance and operational concepts to reduce potential risks. In particular, major advances would be required in life support for the in-space and SHABs; radiation protection and countermeasure development; space transportation propulsion advancements; EDL of large payloads on Mars, cryogenic fluid management; utilization of locally produced consumables, and power systems, to name a few.

A number of technology requirements apply to more than one mission element. These include substantial increases in ground and surface operations automation; in-space system autonomy; and diverse applications of advanced electro-mechanical manipulator systems, using control approaches ranging from teleoperation through telerobotics to full robotics. There would be requirements for data and control system components and software that increase the fault tolerance of system operations, including automated fault detection, isolation, and resolution. Across all phases, human safety and health during long-duration missions would have high priority, and would pace and direct research and technology development.

A major challenge of human exploration of Mars is the need to dramatically decrease the total mass that must be launched into LEO and transported to the martian surface. Although additional factors, such as crew time, power, and servicing requirements, are very important, reducing launched mass is an overarching need for long-term self-sufficiency and acceptable operations costs. Critical technologies in regenerative life support, aeroassist, EDL, and advanced space-based cryogenic engines must be developed to substantially reduce the mass of near-term systems. Mid-term technologies that are critical to decreased mass are surface nuclear power, ISRU, and radiation shielding. In addition, although human expeditions to Mars could be conducted using cryogenic propulsion and aerocapture, nuclear propulsion presents a compelling prospect for tremendously reducing the mass or travel time required.

7.1 Human Health and Performance

As humans extend their reach beyond LEO to the surface of Mars, they will be exposed to the hazardous environment of deep space for lengthy periods; consequently, protective measures must be devised to ensure crew health and maximize mission success. The health and safety of crew members while they travel to and from the Mars and inhabit its surface are key near-term concerns. The explorers must be protected from the space radiation environment and from the physiological effects of reduced gravity. To maintain the fitness and productivity of the crew, medical care must be provided during long stays in very isolated and distant places.

A thorough ground-based research program that is coupled with flight research on the ISS and the lunar surface must be conducted to provide an understanding of the physiological basis for human responses, develop appropriate treatments and countermeasures, and decide how best to support crew members. Simulating the environment that would be inhabited by crews on Mars is an important facet of the research program. Much of the work could be done on the ground, but many studies would require access to space facilities. The ISS is currently serving as a vital test facility for research that demands long exposures to the reduced-gravity loading conditions in spacecraft and on planetary surfaces. That research will establish the baseline for the 6-month transit from Earth to Mars, and is forming the foundation of the extrapolations and inferences that are necessary for near-term planning for the 18-month Mars surface habitation and the 6-month return transit to Earth. In preparation for the Mars missions, research on the moon would be essential. Human adaptation to long-term exposure to partial-gravity conditions is a critical component of future long-duration surface operations on Mars. The moon provides an ideal venue in which to verify and refine the protocols that are established on ISS, as applied to planetary surface operations.

7.1.1 Radiation protection

Protecting the crew from the harsh radiation environment of deep-space travel is vital for a safe human exploration program. Radiation protection goals fall into the following three categories: (1) determination of acceptable exploration career exposure limits and development of countermeasures that could reduce the adverse effects of radiation exposure; (2) provision of sufficient radiation protection strategies in planetary habitats and in the MTV to protect both crews and sensitive equipment from the normal galactic cosmic-radiation background; and (3) establishment of space weather forecasting systems and implementation of sufficient "storm shelters" to warn and protect crews against the transitory, but potentially extreme, levels of radiation that are encountered during solar flares.

The National Council on Radiation Protection (NCRP) guides NASA on crew-permissible exposure limits. However, these limits are based on the characteristics of the radiation that is present in LEO, where Earth's magnetic field provides protection from solar flares and galactic cosmic radiation. Further research is needed to adequately quantify and reduce the uncertainties that are associated with estimating human exposure risks for long-duration deep space and Mars surface radiation environments. For missions beyond LEO, new standards most likely would be established that take into account the inherent risk of exploration missions. The exploration strategy will provide a system of radiation protection that adheres to the so-called ALARA (as low as reasonably achievable) principle, which recognizes that although an acceptable upper limit of exposure is set, the residual risks should be minimized even further where it is reasonable to do so.

Advances are required in radiation protection and countermeasure development from galactic cosmic radiation, including the generation of the secondary radiation that is produced by the galactic cosmic radiation interaction with spacecraft materials. On the Mars surface, the planet's bulk shields against half of the cosmic radiation that is received in space; but again, the generation of secondary radiation from the atmosphere and surface materials may prove to be problematic. Thus, further characterization of the radiation environment of Mars is required. To develop the best radiation shielding strategies for Mars habitats and the transportation vehicles, robotic missions will help to determine the nominal background radiation that is encountered during transits to and from Mars, as well as on the planetary surfaces, and will measure radiation doses that are received during solar particle events (SPEs).

The most acute source of space ionizing radiation for Mars explorers is an SPE, which accompanies some solar flares. The amount of radiation could be so large that the dose to explorers, if unprotected, could significantly exceed all limits and potentially result in rapid death. However, to protect the explorers, "storm shelters" could be constructed in the most heavily shielded areas of the spacecraft and habitats, and could be provisioned with sufficient consumables to maintain humans during the most intense portions (highest dose rates) of an SPE (from a few hours to several days).

A system for alerting the crew is essential to planning EVA traverses, which would not be scheduled for periods in which a flare was expected. Warning must be received in sufficient time to allow the crew to return to the habitat or storm shelter before the buildup of radiation from an unanticipated flare puts the crew at risk. Solar flares are currently unpredictable to the extent that warning times at a spacecraft may be as short as 30 minutes. Improved predictions will require long-term observations of the magnetic field of the sun and its relationship to solar flares, and specific warning systems will need to be developed.

7.1.2 Reduced-gravity countermeasures

Space flight experience has shown that significant physiological changes occur during exposure to reduced gravity; most notably bone mineral loss and architectural changes, muscle atrophy, and cardiac de-conditioning, all of which become more severe without proper countermeasures as the stay-time in space increases. Although these effects could be minimized if crews take certain preventive measures while in space, the problem of developing effective countermeasures to reduced gravity is significant.

The major concern relates to the long transit times to Mars coupled with the demands placed on the crews immediately upon arrival at the martian surface. The baseline transit time to and from Mars is 200 days in zero gravity. Exercise, nutrition, and pharmaceutical countermeasures show promise in controlling the adverse physiological effects of long-duration exposure to reduced gravity. Also, three Mars transit options exist: (1) shorten the outbound and return transit times by using advanced propulsion systems, (2) employ artificial gravity countermeasures within the spacecraft either by providing an on-board centrifuge or by spinning the spacecraft itself, or (3) accept the higher risk involved and proceed with the mission using the best available countermeasures. A zero-g countermeasures program is being conducted on the ISS. Appropriate crew stay-time in orbit, combined with the increase in crew size to six, provides an adequate statistical basis for this vital countermeasure information. In addition, countermeasures that are developed to mitigate the deconditioning effects of microgravity would be used at the lunar outpost and on Mars to maintain crew health and performance in these reduced-gravity environments. Zero-g countermeasures alone may not be sufficient to maintain crew health and performance for a Mars mission, however. Adverse physiological changes due to reduced gravity may be prevented by exposure to some level of artificial gravity, but the specific level of gravity and the minimum effective duration of the exposure that is necessary to prevent deconditioning are not yet known. Although artificial gravity should reduce or eliminate the worst deconditioning effects of living in zero gravity, rotating environments frequently cause undesirable side effects, including disorientation, nausea, fatigue, and disturbances in mood and sleep patterns. If artificial gravity is to be employed, significant research must be done to determine appropriate rotation rates and durations for any artificial gravity countermeasures. The decision on whether artificial gravity must be employed to adequately support crews on their transits to and from Mars, as well as the decision on the necessary gravity level and rotation rate, has significant implications for vehicle design and operations.

7.1.3 Medical care

Maintaining crew health is required to sustain a high level of performance and productivity both in transit and on the surface of a planet. Health systems would be required to provide appropriate medical care, environmental monitoring and regulation, and optimization of human performance. The approach to health and performance systems is to evolve toward increasingly higher levels of self-sufficiency.

Human Mars missions would have small, highly autonomous crews that are situated in remote locations and unable to return immediately to Earth in the event of a medical emergency. Therefore, on-site medical care would be needed to accommodate major and minor illnesses and injuries and perhaps surgical capability. Medical care systems for Mars outposts and for the MTV would build on and expand the capabilities of the ISS to include ambulatory (outpatient) and dental care, emergency medical and trauma response, and advanced life support. These systems would also provide enhanced Earth-based support systems, in-space support, medical computer-aided artificial intelligence systems, and Earth-to-remote locations telemedicine capabilities that use state-of-the-art telecommunication systems for consultation in diagnosis and treatment. As the time that is spent on planetary surfaces increases, medical care capability would also expand to likely provide diagnosis, laboratory analytical capabilities, anesthesia, surgery, and pharmaceutical support. The area of medical care would provide perhaps the most stringent demonstration of crew autonomy, given the 3- to 20-minute time lags in communications with Earth coupled with the potential for urgent life-or-death decisions and actions. One unique issue that is related to medical support on Mars is that the martian surface material itself may present a health hazard to the crew. Analysis for toxic and irritating substances and for any potential biological hazard must be done prior to human exploration; the MSR mission is planned to directly address this issue. Research and advanced development to extend the shelf-life of certain pharmaceuticals and blood products and to develop medical countermeasures against the biological effects of radiation and other oxidative stressors (e.g., dust), are required, as are the development and testing of operational procedures in reduced-gravity and zero-g environments. The capabilities would be demonstrated using moon and Mars advanced development health maintenance facility test-beds.

7.1.4 Supporting human life

Maintaining a safe environment for human habitation goes beyond the minimum required to sustain life by providing adequate air, food, H_2O, and waste handling systems. The habitable environment must also be conducive to maximizing crew productivity by minimizing physiological and behavioral stresses. The environment must be monitored and controlled for the presence of toxins, either microbial or physiochemical, and it must maintain appropriate temperature, humidity, and atmospheric composition. Analytical methods need to be refined to predict the toxic buildup rates and projected levels and sources for extended lunar and Mars missions. Emphasis must also be placed on minimizing the release of gases and biological substances into the martian environment.

Advanced life support systems must have capabilities for air revitalization, H_2O purification, food supply, waste processing, environmental monitoring and contamination control, thermal and humidity control, and fire suppression. Optimizing systems for operation in microgravity during the transit to and from Mars and in the one-third gravity field on the martian surface will need to be traded with the benefits that are potentially gained through commonality in reduced development costs and increased system redundancy

Perhaps the biggest challenge would be the development of regenerative life support systems that minimize the quantity of supplies that must be transported from Earth. Considerations for the design of life support systems for human exploration missions typically lead to the use of closed-loop systems wherever practicable to reduce logistics requirements and to the use of open-loop systems wherever logistics penalties are tolerable within the mission architecture. However, little is known about the operational characteristics and risks that are associated with long-term operation of regenerative life support, especially biologically based regenerative systems. The ISS Program is developing systems that will use recycling technology to supply a portion of the potable H_2O and part of the hygiene and wash H_2O for the crew. For the MTV, development of a more highly regenerative system, which is a natural evolution from the ISS systems but with higher degree of H_2O recovery and less dependence on expendable usage, might enhance mission design while offering the potential for the high reliability founded in years of ISS operational experience. For planetary surface facilities, most of the development work could be ground-based.

In addition, the atmosphere of Mars could be used to generate user consumables to supplement potential losses from the habitats due to outgassing and EVAs as well as to generate O_2 for the crew for enhanced exploration for EVA systems. Oxygen, as well as buffer gases including N_2 and Ar, could be extracted from the atmosphere, in addition to regolith or planetary H_2O, if accessible. Although this is a very promising capability, further assessments, including the human health impacts of such an approach, must be fully developed.

7.1.5 Behavior, performance, and human factors

Space environments, like other isolated and confined environments, induce stress as well as physiologically and psychologically induced stresses. If not appropriately managed, combined stress is likely to result in behavior and performance deterioration during long-duration space missions. Humans have never embarked upon space flight missions approaching the scale of exploration now envisioned; the best analogs so far may be Antarctic and undersea expeditions. Although these Earth-based analogs are not perfect, they provide insight into some of the unique attributes that are present in space exploration – e.g., alteration of day-night cycles, telecommunications to outpost operations, absence of other living creatures, self-sufficiency, and profound isolation – that affect crew dynamics and performance on space exploration missions. Using the analog environments and specialized ground simulation facilities as test-beds, and building on data obtained from the ISS and the lunar outpost, strategies will be developed to support the increasingly complex and demanding Mars missions.

The exploration missions must be carefully examined from a space human factors perspective. The key issue is the effect that prolonged exposure to the space flight environment has on individual psychological and behavioral functioning and on crew effectiveness and performance. Spacecraft architecture and outfitting are particularly relevant based on their impact to psychological, social, perceptual, and behavioral conditions affecting crew performance, productivity, and safety.

Crew composition would be based on personal and interpersonal characteristics that promote smooth-functioning and productive groups, as well as on the skill mix that is needed to sustain complex operations. Studies need to be conducted that address these areas, as well as the influence of task and authority structures and the introduction of new members and unfamiliar crews, to determine effects on crew performance and productivity. Crew composition, training, and skill mix and retention should be studied in ground-based laboratories, simulations, and analog test-beds.

7.2 Space Transportation

7.2.1 Earth-to-orbit transportation

The Ares V HLLV, which is currently under development in NASA's CxP, is an enabling capability for placing payloads of large mass and volume into LEO. The Ares V lunar reference vehicle is specifically designed for minimum technology development while meeting the requirements of the architecture. Some challenges, however, will inevitably arise out of the sheer physical scale of the Ares V, which in many respects surpasses that of the Saturn V lunar rocket. The Mars payloads to be integrated would be factors of two to four times larger than any previously attempted. Ground handling and operations of these payloads would need to be carefully studied and understood. Support operations would likewise be challenging in most aspects, as the Ares V would require more fuel (cryogenic), support gasses (helium (He)), transportation needs, and larger boosters than have previously been supplied, which would likely lead to a more complex support integration process. The Mars launch campaign, which would require multiple launches, would need a much higher rate of launch than has been attempted with vehicles of this magnitude and would also require very long duration loiter capabilities for the payloads. Composite structures figure prominently in the Ares V launch vehicles, including the primary structure and payload shroud. In addition, options are being studied to improve performance through the incorporation of composite SRB cases and changing to HTPB (hydroxy-terminated polybutadiene) propellant, which is more energetic and could create a higher internal pressure, thereby providing substantial performance improvement.

Another key driving challenge for the launch vehicle in support of future Mars missions is the capability to not only lift large payloads, but also to lift large volumes. A packaging and support strategy of using the launch vehicle payload shroud as the primary structure for the Mars EDL system is very promising. Further assessments in terms of the ability to integrate the Mars lander, the appropriate shape of the launch vehicle shroud, and TPS approaches are necessary to determine the feasibility of this approach.

7.2.2 Advanced chemical propulsion

The technology that is proposed for the DRA 5.0 Mars descent and ascent propulsion systems uses a pump-fed LOX/CH_4 propellant combination. The O_2-based propulsion system was chosen not only to improve performance, but also so that ISRU could be used to produce the required ascent O_2 at Mars instead of having to carry it from Earth. This is not a new concept, having been thoroughly analyzed in previous Mars Design Reference missions. Currently, no pump-fed LOX/CH_4 engines are in production, and only pressure-fed engines are in development. Much research and testing is required to produce a highly reliable, pump-fed engine that could meet the human Mars mission requirements. The LOX/CH_4 engines face the challenge of having to start after sitting idle for an extended period of time (in this case, on the martian surface). Pressure-fed engines have been considered to alleviate this concern. Without the rotating turbo machinery, pressure-fed engines are much simpler and more reliable than their pump-fed counterparts. However, while the engine dry mass for the pressure-fed engine is lower than for the pump-fed engine, the overall feed system mass is much higher due to the higher pressure that must be maintained in the propellant tanks (250 psia vs. 50 psia). The required helium pressurant (and tanks) is also greater. This problem is made even worse due to the lower (in general) Isp, and corresponding higher propellant requirement, of the pressure-fed technology. These factors result in much lower payloads that could be delivered. Therefore, pump-fed engines are chosen for CH_4 engines in the current mission. Also, the work and testing that is required to verify that the LOX/LH_2 engine starts after long idle times would, hopefully, solve any issues with the LOX/CH_4 start capability.

7.2.3 Nuclear thermal propulsion

NTP is a proven technology that has the potential to enable future human Mars missions with reasonable mass requirements and credible numbers of Ares V launches. The technology was demonstrated to high technology readiness levels (TRLs) during the Rover/NERVA programs. A variety of fuel forms were developed, and a broad range of different thrust-class engines were ground-tested at the Nevada test site. Near the end of the program "open-air" testing of engines was replaced by "contained" testing, using an effluent treatment system to process the H_2 exhaust. While the continued development of chemical propulsion systems has led to performance advances in the non-nuclear engine and stage component areas (e.g., H_2 turbopumps, regenerative-cooled nozzles, and lightweight cryogenic tanks) that are required for the NTP MTV, further work and funding is needed in the nuclear area that is aimed at achieving the following objectives: (1) establishing firm NTP engine system requirements using updated Mars mission analysis and payload estimates; (2) recapturing "composite" Rover/NERVA fuel element technology, and maturing uranium dioxide UO$_2$ in tungsten metal "cermet" fuel technology; (3) performing high-fidelity modeling, design, and engineering of candidate engine systems; (4) preparing the necessary test facilities; and (5) conducting the required nuclear/non-nuclear demonstration tests of NTP fuels, components, and subsystems in preparation for "contained" full-scale ground-testing of demonstration engines and, after this, flight-type engines.

7.2.4 Entry, descent, and landing

NASA's current ability to land robotic payloads on the surface of Mars is largely reliant on the EDL technology set that was developed during the Mars Viking Program in the late 1960s and early 1970s. NASA's flagship 2011 Mars mission, the MSL, has reached the landed payload mass limit capability (approximately 1 t) using the Viking-based technology set, which includes the blunt body 70-degree sphere cone aeroshell, the super lightweight ablator (SLA) 561-V TPS material, and the supersonic disk-gap-band parachute system. The 1-t landed mass capability of the MSL EDL system is a factor of 40 below what would likely be required to achieve a human-scale Mars mission. As NASA strives to land larger mass robotic missions – e.g., MSR – and looks forward to human missions to Mars, additional EDL technologies must be identified and developed to the point that they become viable candidates for robotic and/or human mission sets. In addition, technologies that would be adequate for the lower range of masses may not be applicable or scalable to the very large landed mass EDL systems that are associated with human-scale missions. Because of the limited NASA technology budget, it is imperative that the minimum-cost technologies, which are required for the entire range of desired landed masses that provide the required reliability, be identified.

The very low atmospheric density at Mars prevents the use of traditional terrestrial aerodynamic decelerators as a means by which to attain subsonic velocities for landing as is done on Earth. The challenges that are associated with the development of a human-rated high mass (100+ t) Mars entry system remain large. While there is considerable uncertainty in the ultimate outcome of human-scale landing system designs, several technology options provide candidate pathways. Certain combinations of aerocapture and entry, descent, and landing (AEDL) technologies that may be achievable and that may result in robust performance and acceptable risk architectures have been identified and deserve further study. Several of these options under consideration include slender body systems, blunt body systems, supersonic aerodynamic inflatable decelerators, supersonic retro-propulsion, rigid hypersonic deployed decelerators, and hypersonic inflatable decelerators (e.g., inflatable heatshields).

Advances in TPS technologies would also require significant investment and be largely dependent upon the AEDL architecture that is selected for Mars. These TPS technology development efforts, which are integral to the technology and system-level development and down-select process, need to be conducted as parallel development paths. In addition, since it is apparent that no off-the-shelf ablator for a block upgrade of the Orion, even at 12 km/s, exists or will be forthcoming from the lunar Orion vehicle, a large technology gap in TPS exists, thus leading to the requirement for research and development on TPS materials. As NASA moves into full development of the Orion vehicles, more attention should be focused on addressing the needs of Orion upgrades and also on ablators for delivery of heavy mass payloads to the surface of Mars.

7.3 Surface Systems
7.3.1 Advanced habitation systems

Structural materials advancements that would provide large livable volumes, both in transit to and from Mars, as well as during surface exploration, while minimizing mass are desired for human exploration missions. Limited volumes and the complexity of packaging the Mars lander and surface systems within the aerodynamic shell of the entry system would most likely require advanced inflatable structures. Key technology thrusts include habitat concepts and emplacement methods (including remote and autonomous operations) as well as advanced lightweight structures (inflatable vs. traditional "hard shell"), and developing integrated radiation protection for crew health and safety. In addition, developing technologies that could significantly reduce the consumables that are required to support the crew during long-duration missions are also critical for the human exploration of Mars. Technologies include air and H$_2$O loop closure, environmental monitoring, solid waste processing, thermal control, and food processing. Advanced sensor technologies to monitor ,and intelligent systems to control, the environmental "health" of the advanced life support system, including air and H$_2$O, are needed. Advanced habitation systems must be easy to maintain, repair, and operate in light of the limited resupply and logistics capabilities. These factors lead to the need to provide the crew with the ability to conduct repair at the lowest level of component possible.

7.3.2 Extravehicular activity and surface mobility

The success of exploration missions depends on the ability of humans to work on and explore planetary surfaces. This success would depend on productive EVA conducted at great distances from the surface landing site or outpost. During these missions, astronauts would be exposed to a range of gravity conditions and a diversity of environments. With the normally intense activity expected on the exploration missions, issues of productivity, usability, durability, and maintainability of EVA systems become acute. Operational and medical considerations would include pre-breathing procedures, life-sustaining system capability, environmental health, radiation protection, and emergency-mode operations. Allowing humans to make the transition simply and effectively between activities inside and outside vehicles would both enhance productivity and increase overall mission safety. EVA systems must be provided for the moon, Mars, and space operations in orbit and in transit.

Planetary surface systems, including spacesuits, must be lightweight, would have to be maintained by the crew, and must be resistant to contamination by surface materials such as dust. EVA systems must provide a safe, nontoxic environment, with food and H_2O supplies that are nutritious, esthetically pleasing, and free of contamination. It is important to minimize restrictions on human capability by providing adequate thermal control, greater suit mobility – in particular in the gloves, torso, and boots – and enhanced communications capability for explorers and home base interactions. Maintainability of the system, allowing reuse without extensive overhaul, is also vital for a robust exploration strategy. Suit development is an area that would require focused research and technology efforts emphasizing lightweight and durable materials, glove design, dust contamination protective measures and techniques, lower torso mobility systems for walking, ancillary mobility systems for surface transportation, long-term reusability and lightweight, compact, PLSS technologies. Although the lunar surface has the potential to increase the EVA experience that would be needed for crewed Mars exploration, a number of differences between the martian environment and the lunar environment would dictate changes in the EVA architecture. These environmental differences include the increase in martian gravity to 0.38 g relative to 0.17 g on the moon, and the difference between the lunar atmospheric "pressure" of 1.3×10^{-13} mbar and the 10-mbar atmospheric pressure on the martian surface.

Unlike the lunar surface, exploration of Mars must be conducted with issues relative to planetary protection in mind, particularly management of organic contaminants that are released by EVA suits and PLSSs. The nature of pressure garments is that leakage of some quantity of internal atmosphere is unavoidable. Leakage includes the gas that was used to pressurize the garment as well as any airborne particulates that could escape past garment seals, which could include microbes and latent virus particles that are shed from the crew members. A critical engineering and operational challenge will be to manage this leak rate, potentially through the use of improved seals, sterile overgarments, or covers around mechanical connection areas such as glove and neck rings and through the use of operational practices that minimize human crew member entry into areas that are suspected of having extant or fossil martian life.

For Mars surface exploration, scientific diversity is obtained by extending the range of human explorers via both unpressurized and pressurized rovers. Long-range pressurized rovers may be large, complex machinery upon which much depends. A thorough understanding of operational issues and failure modes will be essential. Both Earth analogs as well as lunar missions would play a vital role in determining the performance and operational scenarios, including maintenance and repair, of surface mobility systems. Since surface mobility systems would use many of the same types of mechanical equipment, structural elements, and materials as other mechanical surface systems, long-term reliability of large rover systems in extreme planetary environments needs to be established. Common systems need to be qualified for multiple uses. Demonstration of systems performance in harsh environments, such as testing done by the US Army's Cold Regions Lab and data from MER and other future Mars rovers, is highly relevant. We need to learn how to make mechanical elements perform reliably in environments for which we have little direct experience.

7.3.3 Subsurface access

Science requirements for the mission call for both shallow (tens of meters) and deep (hundreds of meters) drilling and collection of samples from the subsurface of Mars. Research on the cuttings transport method is probably the most important element of design that must be addressed, first theoretically and then in the terrestrial laboratory, before selecting the best approach for Mars. Cuttings transport simply would not behave the same way as it does on Earth due to the low gravity and low atmospheric pressure and temperature on Mars. Efficient cuttings transport would be important, especially since rapid rate of penetration is desired. For the shallow drilling case, dry rotary drilling with augering of cuttings may suffice but is likely to be slower than desired. Alternatively, a pressurized stream of martian atmosphere could be used to aid in cuttings transport that is produced on a drill site using an air compressor. This method may also work in a deep hole to assist with transporting cuttings into a cuttings cup that is periodically shuttled to the surface and emptied when full. However, for deep drilling, the use of low-temperature drilling mud that is based on H_2O/brine combinations could plausibly work better, so this approach should be seriously evaluated.

Research on bit design for Mars is also needed. Cutters need to be chosen that are optimized for either a wide range of materials that might be encountered, or a selection of bits provided that are specialized to the most likely materials that would be encountered with bit changeout to occur according to the material type that is actually encountered. Provisions should be made for bit changeout due to bit wear every 10 to 20 m of drilling. This means that the bits must be considered an expendable resource and bit changeout, either by autonomous means or assisted by the crew, should be planned for.

For shallow drilling, it is feasible to retrieve the drill string and core barrel to the surface for each core retained. For deep drilling, this would be time prohibitive, so a core barrel that could be shuttled to the surface without retrieving the drill string would be necessary. Deep drilling would almost certainly require hole casing. Casing would be used to stabilize the hole, thereby preventing the collapse that results in getting the down-hole elements stuck and produces the loss of the drill head and/or string. Studies should identify the most mass and volume-efficient casing material, such as inflatable tubing anchored by flexible cements, for sealing the casing into the borehole.

As with EVA systems, subsurface exploration of Mars must be conducted by keeping issues that are relative to planetary protection in mind. Current understanding of the martian environment indicates that extant life may be more viable in subsurface areas where temperatures, radiation levels, and potential access to liquid H_2O are closer to those conditions where life is known to exist and thrive. Under these circumstances, subsurface regions must be considered "special regions," as defined earlier, until proven otherwise. This indicates that technology would be needed to prevent the introduction of contaminants from the crew or its equipment as well as detect evidence of present or past indigenous life. Appropriate sample handling technology and protocols would also be needed in the event that signs of life are detected – a primary goal of Mars exploration.

7.3.4 Nuclear power generation

Providing robust continuous surface power is critical for future exploration of the martian surface. Fission surface power (FSP) technology advancements would build on previous activities while expanding the breadth to include reactor and shield-related development. Additional component technologies that could be pursued specific to the FSP application include reactor fuels, structural materials, primary loop components, shield materials, high-power Stirling conversion, and high-voltage power management and distribution. On the nuclear side, initial irradiation tests could be performed on candidate fuel forms. In parallel, materials testing could evaluate radiation effects and fill gaps in thermal-mechanical property databases. Additional reactor-related items for development include primary pumps, heat exchangers, accumulators, control drive actuators, and instrumentation. Since shielding exercises a major influence on design and mass, several early experiments could be conducted to evaluate material and packaging options. On the plant side, component development activities could expand on previous efforts while focusing on lunar and Mars environment issues. Of particular interest would be radiators and transmission cabling that are suitable for planetary surfaces and amenable to the various power conversion options. The component technology element would also include the further development of multi-kilowatt, 900 K Stirling converters.

7.3.5 Solar/regenerative fuel cell power systems

A PV solar power system uses solar cells that are configured into an array and typically coupled to an energy storage device such as a fuel cell. Energy storage is required to provide power when the array does not see the sun or when power output is attenuated below load requirements. Energy storage also answers peak power demand. Current solar cells that are available and achieve 27% energy storage include, for example, the advanced triple junction GaAs/Ge (gallium arsenide/Germanium), which are the cells that are used on the MERs. Even with high-efficiency solar cells, array areas that are needed to produce the required power for a human mission become very large. For solar systems to be competitive at Mars, advances in cell efficiency, dust mitigation, array deployment, and operational maintenance strategies must be improved. Since the power system is pre-deployed prior to the arrival of the crew, a robust method of robotic or autonomous deployment, anchoring, checkout, and operation of large array systems must be developed. The Mars DRA study array option was 2.5 m high × 58 m long, the total system of which is comprised of 10 array wings.

The environment of Mars, which is rife with dust accumulation and dust storms, would profoundly affect the overall performance of solar power generation systems. Previous robotic missions to the surface of Mars have provided valuable data with respect to dust accumulation. A robust method must be identified that could operate robotically since the arrays must be operational prior to crew arrival. Technologies for dust mitigation, such as compressed gas "blow off," mechanical wiping, vibration to fluff off the dust, and electrostatic repulsion, have been considered; and further work is required to determine the best approach, particularly for large arrays. The MER has shown that random wind events could occur and may, in some circumstances, restore lost power generation due to dust accumulation. But these dust-clearing events are random, very localized, and, thus, could not be expected to occur. Active dust mitigation approaches must therefore be developed and incorporated in future solar system designs.

Advanced energy storage devices are necessary to supply the necessary power for nighttime operations and during dust storms. Advances in both primary fuel cell power systems and RFC energy storage systems are being pursued. An RFC system is a combination of a primary fuel cell and an electrolysis system, along with associated integration hardware. The fuel cell and RFC work is categorized into six major areas: (1) flow-through primary proton exchange membrane fuel cell (PEMFC) development, (2) non-flow-through primary PEMFC development, (3) high-pressure electrolysis development, (4) RFC technology development, (5) passive thermal development, and (6) advanced membrane-electrode-assembly (MEA) development.

7.3.6 Isotope power systems

Isotopic power systems offer continuous power much like the nuclear fission system. Their practical range is on the order of several kilowatts due to the availability of ^{238}Pu, which is produced by Neptunium-237 (^{237}Np) neutron exposure. ^{238}Pu has many attractive features compared to other isotopes, lower radiation (minimal, low-mass shadow shield), high-power density, and an 87.7-year half-life. ^{238}Pu has fueled all of the RTGs that are used in NASA missions.

NASA's use of radioisotopes is well established since Apollo (the Apollo lunar surface experiments package (ALSEP)) and has enabled over 30 outer planet missions as well as the Viking landers. These systems work by converting the natural radioactive decay heat (largely alpha particles) into an electric current. The thermoelectric devices are limited in conversion efficiency; thus, high-power systems would require large amounts of radioisotope fuel. The Savannah River facility, which produced the ^{238}Pu, has been shut down with plans to restart production at a combination of alternate facilities at future date. There is currently a limited supply of ^{238}Pu and a strong competition for it to support future NASA missions.

Advances in power conversion, such as Stirling generators, are needed to improve the efficiency of converting thermal heat into electrical power. The advanced conversion technologies that are proposed could provide a four- to five-fold increase in isotope utilization, thus drastically reducing mission cost while making prudent use of our scarce resource of isotope fuel for future missions.

7.4 Cross-Cutting Systems and Miscellaneous Needs
7.4.1 In-situ resource utilization

The use of non-terrestrial resources could provide substantial benefits to a variety of future space activities by dramatically reducing the amount of material that must be transported from Earth to a planetary surface. ISRU is a critical component of long-term, largely self-sufficient outpost operations. By extracting and processing local resources to obtain or make O_2, H_2O, CH_4, and buffer gas consumables for life support, EVAs, and ascent propulsion, significant mass reductions or increased payload to the Mars surface is possible. There are two primary resources of interest on Mars: (1) the atmosphere, which is mostly made up of CO_2 (95.5%), N_2 (2.7%), and Ar (1.6%); and (2) the H_2O that exists in the top meter of Mars soil. Since NASA and international robotic missions have shown that H_2O can be found globally across the Mars surface, and Mars Odyssey mission data suggest that there are regions with up to 8 to 10% H_2O by mass in the top 1 m, the extraction and use of Mars H_2O raises both benefits as well as challenges to future missions.

Before Mars atmospheric CO_2 could be used or processed, it must be collected, separated, and pressurized; typically at or above Earth ambient pressure (>14.7 psia) to increase the efficiency of CO_2 processing concepts. Methods for CO_2 collection and pressurization include: mechanical pumps, micro-channel adsorption, and cryogenic separation (CO_2 freezing). Since mechanical pumps have been shown to be ineffective in compression of the martian atmosphere to the desired levels, advances in micro-channel adsorption and cryogenic separation are necessary.

Conversion of atmospheric CO_2 into O_2 could be performed in a number of different ways, depending on the resources that are available and the products that are desired. The three processes that have been examined the most due to process simplicity or commonality with life support systems are: (1) CO_2 electrolysis, (2) Sabatier conversion of CO_2 to CH_4 and H_2O (with subsequent H_2O electrolysis), and (3) reverse water gas shift (RWGS) conversion of CO_2 to CO and H_2O (with subsequent H_2O electrolysis). For both Sabatier and RWGS conversion of CO_2, H_2 is required. In the case of O_2 production using RWGS, the H_2 that is required is obtained from the subsequent H_2O electrolysis, so H_2 is recycled. In the case of O_2 production using Sabatier, only half of the H_2 that is needed is recovered from the subsequent water electrolysis process. It should be noted that while other technologies and methods for CO_2 processing are possible and have been evaluated, these processes were considered to be too low of a TRL to be evaluated at a system level for mission applicability. These alternative, low-TRL technologies include: molten carbonate electrolysis, non-aqueous electrolysis of CO_2, ionic liquid electrolysis, liquid CO_2 electrolysis, and lower-temperature mobile oxide ceramics. Finding H_2O in sufficient amounts, and of sufficient accessibility, provides an alternative to these low-TRL technologies, via use of electrolysis.

The combined Mars atmosphere and soil/H_2O processing option that is based on the use of a Sabatier reactor and H_2O electrolysis combined with an excavator/rover and a soil processing reactor was found to be very attractive. The concept is based on combining past Mars ISRU designs with current work that is being performed to model and develop lunar regolith excavation and regolith processing systems to extract O_2 from regolith. To extract H_2O from Mars soil, the soil is heated to approximately 600 K, and an inert gas flow fluidizes the soil to help desorption of H_2O. The inert-H_2O gas stream is sent to a gas clean-up process to remove any contaminants that were created during the process, and the H_2O is than collected and electrolyzed to produce O_2, which is liquefied and stored, and H_2, which is sent to a Sabatier reactor with Mars CO_2, to make CH_4 and more H_2O. The Sabatier/H_2O electrolysis process is very similar to the system that is used for habitat life support processing, and the soil/H_2O processing is similar to lunar regolith O_2 extraction processing, where lunar regolith is fluidized and heated to 1270 K with H_2 to produce H_2O from iron (Fe)-bearing minerals. Even with strong cross-cutting ties to life support and lunar ISRU, limited concept evaluation to date and Mars surface water property and distribution uncertainty would not allow this process to be baselined at this time. It is believed that this approach should continued to be evaluated in light of current and future Mars robotic missions that will examine soil properties and H_2O on Mars (such as the current Phoenix mission) and lunar excavation and regolith processing technology and system development.

7.4.2 Cryogenic fluid management

Cryogenic fluid management (CFM) is a critical technical area that is needed for the successful development of the Mars architectures. The first and foremost challenge is the storability of LH_2, CH_4, and O_2 propellants for long durations. Note that the longest flight of stored cyrogens is Titan Centaur-5, where the propellants were stored in orbit for a 9 hours. These propellants have very low boiling points – well below the environment temperatures of Earth orbit, Mars transit, Mars orbit, or Mars surface – as such, the tanks must be regularly vented to prevent over pressurization. Such venting would cause unacceptable propellant losses for the long-duration missions to Mars that are being considered. In lieu of venting, active cooling or refrigeration could be integrated to the tanks to preserve propellants. Most aspects of long-term cryogenic storage technology exist at some state, mainly from the development of advanced dewars for life support and satellite instrument purposes. Thick multilayer insulation (MLI) systems have been applied to cryogenic dewars; also, active cooling components, such as cryocoolers, have been integrated. These are rapidly advancing in capability and state of the art and are gradually replacing cryogenic dewars for space telescope applications. Nevertheless, these developments have not been applied to cryogenic propellant applications, particularly to the size of tanks that are needed for this Mars architecture. Furthermore, there have been no significant advances in LH_2 temperature cryocoolers near the sizes needed for zero boil-off cryogen storage.

Besides the thermal control aspects, other CFM development issues that would ensure safe and reliable cryogenic storage and supply to the propulsion systems include liquid acquisition and transfer, to ensure vapor-free propellant supply to the engine as well as to a second tank, and mass gauging, to ensure reliable propellant quantity information. These three cryogenic areas have been under development for the present CFM program, which is part of the Exploration Technology Development Program (ETDP). The purpose of that effort is to mitigate the substantial risks that are associated with cryogenic propellants by the year 2011, in support of lunar mission architectures. Note that all of the technical elements under development by CFM are applicable to the Mars mission scenarios.

Another system that would benefit from advanced cryogenic propellant storage systems is the ISRU. Advanced storage systems include large-scale, flight-rated cryocooler development, which is central to large-scale liquefaction efforts. Furthermore, this same development would benefit concepts for efficient long-duration storage of fuel cell reactants.

7.4.3 Communication and navigation

Advances in cross-cutting technological information systems and automation have the potential to substantially improve the performance, increase the reliability, and reduce the cost of the systems that are required for future human exploration of Mars. Technology development could decrease delay rates and increase system on-line availability through fault-trend analysis and management. Development could also substantially decrease the operational effect of the need for space-based system maintenance through the application of highly automated system architectures. Technological advancements are needed in automation and robotics, high-rate communications, and data systems, processors, and recorders.

The Mars network that would be required to support human Mars missions would be an extension of the current NASA space networks and the planned lunar network. The fundamental differences between the lunar and Mars architectures can be ascribed to a few factors as follows:

- A 1,000-fold increase in maximum distance (400M km vs. 400K km)
- Mars environment including atmosphere
- Choice of a circular areostationary orbit vs. elliptical, eccentric orbit
- Spectrum requirements for deep-space (Category B) vs. near-Earth (Category A) mission

While Mars missions are expected to use comparable data rates as lunar missions, achieving those levels of performance on the much longer Earth-Mars link would represent a key challenge, requiring significant advances in communications capability.

The larger Earth-Mars distance drives the design to X-band (vs. S-band for near-Earth use) and increases the difficulty of closing the communication link by 1,000,000-fold in proportion to the square of the distance, relative to Earth-Moon links. And, the envisioned human-era data rates represent a growth of roughly two orders of magnitude beyond current state-of-the-art Mars robotic exploration capabilities. Closing this gap would require a combination of increased transmit power; larger, more directive spacecraft antennas; transition to shorter-wavelength Ka-band or, potentially, optical links; and an increased receive aperture at Earth. The martian environment drives the configuration of antennas for descending and ascending vehicles and link budgets through the atmosphere. The areostationary orbit, which is highly advantageous for communications purposes, provides continuous coverage of one-third of the planet with a single asset but sacrifices navigation utility compared to the lunar orbit.

7.4.4 Supportability and maintainability

Among the challenges facing human Mars missions will be the development and implementation of robust supportability concepts. In the current context, the term "supportability" has a rather broad scope that includes system maintenance, maintenance-related processes, maintainability design issues, crew support functions including provisioning and overhead tasks, and other issues that fall within the scope of integrated logistics support. Supportability issues will be so important to mission success that they must be an integral part of the operations concept and, in fact, will be a key factor in the development of hardware design requirements.

Resupply capability for human Mars missions are essentially nonexistent. All resources that are required to support the mission must be pre-positioned or carried with the crew, with the exception of resources that are generated in-situ. These missions would also face mass and volume limits that would restrict sparing options and strategies. These two constraints highlight the need for, and challenge of, a self-sufficient supportability approach. It would be necessary for the crews of these missions to have at hand all of the resources that are necessary to sustain critical spacecraft systems and support equipment for the duration of their time away from Earth. This capability must be provided while minimizing associated mass and volume requirements.

This self-reliance would be achieved, in part, by increasing emphasis on maintenance by repair rather than replacement. A repair-centered maintenance approach would only be effective, however, when it is strategically coupled with hardware design that is specifically structured as part of the supportability concept.

7.5   Risk Mitigation Strategies

The human exploration of Mars would be a complex undertaking. It is an enterprise that would confirm the potential for humans to leave our home planet and make our way outward into the cosmos. Although just a small step on a cosmic scale, it would be a significant one for humans because it would require leaving Earth on long missions with a constrained return capability. The commitment to launch is a commitment to several years away from Earth, and there is a very narrow window within which return is possible. This is the most radical difference between Mars exploration and previous lunar explorations. Successful implementation of human exploration of Mars would require a thorough and in-depth technology development program that is coupled with a rigorous risk mitigation strategy.

Precursor activities consist of risk analysis-based technology investment and test and validation as required for producing the technical readiness to develop human missions to Mars. In addition to leveraging the technical advances that are expected from the lunar human missions and the ongoing robotic Mars Exploration Program (MEP), new unique precursor activities must be initiated to pave the road to the required capability readiness.

Although no specific timetable has been established for the first human mission to Mars, a notional date of the early 2030s was used as an example date for assessments by the Mars architecture study team. For the initiation of Mars human missions in the early 2030s, mission development would be initiated in the 2020 to 2030 time period. Thus, precursor activities would need to be conducted in the mid-2010s to early 2020s via a series of system studies, technology developments, and validation tests, including possible robotic precursor flights to Mars. This sequence may culminate with large-scale precursors early in the third decade to validate design approaches.

7.5.1   International Space Station and near-Earth missions
The MAWG spent most of its effort during 2007 on the applicability of lunar and Mars robotic missions and less of its effort on Earth and ISS testing objectives. Although the time required for a comprehensive assessment was not available during the 2007 study, some preliminary efforts were spent on LEO testing. Missions to the ISS are providing the critical data and knowledge that is necessary for development of zero-g countermeasure protocols. In fact, each mission that is flown to the ISS serves as an essential "Mars transit simulation" since it mimics the 180-day transfers to and from Mars. Each crew that rotates through the ISS provides the fundamental data that are necessary for certification of the zero-g Mars transits. The ISS also serves as a vital test-bed for other critical technology and research areas including radiation monitoring and shielding technologies; advanced medical care; life support system closure and long-term maintenance; automated rendezvous and docking techniques; safe and reliable delivery and return of the crew via the Orion vehicle; certification of low-speed entry guidance and TPSs; as well as key supportability concepts including low-level component repair techniques, long-term system performance, and understanding and validation of gravity-sensitive phenomena.

7.5.2   Lunar missions
Missions to the moon represent the logical first step in exploration beyond LEO, leading to human missions to Mars. The moon is a natural body with reduced gravity (one-sixth that of Earth) and has a total area that is roughly equivalent to the continent of Africa. It is relatively close, only a few days away, and is a natural research laboratory that is orbiting planet Earth. Lunar missions and surface activities are essential activities prior to accomplishing piloted

missions to Mars. The moon is relatively accessible and return could be accomplished at any time, unlike the eventual missions to Mars. This close proximity and enhanced risk posture allows the moon to serve as a vital proving ground for the more difficult Mars missions. The topography and environment of the moon could be used to simulate martian conditions and remote operations. As part of this approach, it is important to test and operate actual equipment and systems that would be used for the Mars missions. The only way to prove that the equipment and systems are truly reliable is to test their functions and operate them over long periods of time in realistic environments.

The moon provides a realistic testing environment for human performance that would yield the understanding that is needed to ensure the safety of the crew. The issue of human performance after long exposure to zero g, and the effectiveness of countermeasures to long-term exposure to zero and reduced gravity, must be well understood before sending crews to Mars. For instance, the degree of autonomy that is required in systems and equipment is better assessed after understanding crew adaptability to reduced-gravity environments. Mars stay times could be simulated by using time spent on the lunar surface and ISS. Crew members' adaptability could be measured in facilities on the moon, while performing tasks similar to those required at Mars. These crew members would also experience the psychological effects and isolation that are experienced by crews traveling to and from Mars. In this way, operational concepts could be developed to make best use of the systems and crew on the planetary surfaces.

In addition, the moon provides a unique location for the testing and operation of similar Mars surface hardware for extended periods to build confidence in reliability and maintenance procedures. This includes items such as life support systems, habitation systems, rovers, power, and ISRU, to name a few. Emphasis should be placed on maximizing commonality between lunar and Mars surface hardware technologies to reduce development and operation risks. Mars operational concepts could be tested on the moon, including those that, at this early stage, could be considered "risky," such as O_2 transfer to EVA suits and rovers or reliance on ISRU for enhanced surface operations.

7.5.3 Mars robotic missions

The need for establishing environmental measurements and testing of technologies and systems at Mars is discussed in several of the previous sections in this report. EDL challenges for human flights to Mars, for instance, require extending today's robotic program capability of landing 1 t of useful mass on the surface to 40 t. Use of in-situ resources used at Mars, both for breathing O_2 and for propellants, present difficult engineering challenges in a foreign, hostile environment. Challenges such as these would require single or multiple test flights to Mars. While some environmental measurements may be undertaken in concert with robotic scientific missions, some test flights would lie beyond the fiscal scope of the robotic science program and must, therefore, be considered as precursor programs in the ESMD.

Recent mission studies have shown that a single Ares V launch to Mars could carry as much as 40 t of mass to the vicinity of Mars. This allows a relatively large-scale instrumented demonstration of aerocapture to orbit and aero-entry to landing resulting in masses between 8 and 12 t on the surface. These heavy landers could host a large variety of tests on the surface of Mars, including subsurface drilling to approximately 10-m depth at several locations via surface mobility as well as, perhaps, quarter-scale ISRU prospecting and productions plants. Many measurements of climate, soil, and dust properties and surface chemical composition could also be performed. Alternatively, if desired, a single Ares V launch could enable a sample return from one or more Mars surface locations. This would allow chemical measurements of surface elements to extreme accuracies by the best instruments on Earth. Such missions to Mars should be planned for no later than the middle of the century's third decade, serving as the last phase of precursors needed to enable human travel in the fourth decade.

The robotic MEP is a science-driven, technology-enabled program of successive missions to Mars. There is much synergy between its drive to gain scientific knowledge of the martian surface and atmosphere and the needs for martian environmental knowledge to enable safe human travel to from the planet's surface and existence on it. Increased knowledge is required in several areas, including atmospheric composition and dynamics on the surface and at higher altitudes, H_2O accessibility, H_2 abundance, neutron and charged-particle fluxes, winds, dust characteristics, soil chemistry and trafficability, biohazards, surface toxicity, and others. Studies have also identified the potential for robotic program contributions in engineering areas, including subscale aerocapture, in-situ resource prospecting and usage, higher-precision landing systems, materials degradation, autonomous orbit rendezvous and navigation, and creation of a continuous communications infrastructure. It could be reasonably anticipated that the robotic program would continue into the second decade and host missions for Mars environmental measurements that could benefit readiness for human travel. It is recommended that the ESMD and the SMD partner on missions that achieve the needed environmental knowledge expansion. A productive partnership here would require funding from both offices.

Because of various factors, however, the masses of robotic science program flight vehicles may be limited to 4 t of entry mass with 2 t of surface-landed mass, which is a factor of about 20 to 40 below the masses that are required by human vehicles. An aggressive program in the ESMD, with substantial partnership funding, would be required to leverage the current robotic science program to make significant progress in the engineering disciplines where relatively large-scale testing is required, such as EDL and in-situ resource production for human missions.

8   Public Participation

The human journey to Mars is a venture in possibility that is sustainable only by the courage, capability, and commitment of people from around the world over many decades. Each robotic and human-precursor mission in a series would build, step by challenging step, the infrastructure for the first three human missions to the red planet. While this dramatic voyage beyond our home planet would stretch technological capabilities, crew health and safety tolerances, time and budget constraints, and the will of hundreds of thousands, we as humans strive to go because Mars has much to tell us about the possibility of life beyond Earth, climate changes on both planets, and the limits of survival and sustainability.

Salute – Active public engagement is a critical element of future human exploration of Mars. Rawlings 1995.

The human journey to Mars does not just refer to the brave select few who would dare to go, or just to the thousands more in the space industry who, over the course of their dedicated careers, would build, launch, and monitor the missions. This voyage belongs to the people of the Earth, who collectively would not only be witnesses to history but could increasingly be participants in it. We are already experiencing the beginning of a virtual human presence on Mars, first through the connection between mission team members and their rovers, landers, and orbiters, which have become robotic extensions of themselves, and second between these robotic partners and the public, who have followed their paths, seeing Mars through their "eyes." Advances in internet and broadband technologies are enabling people not just to access information, but increasingly to have dynamic experiences and make contributions of their own.

Given the large investments that are required and the risks that would be incurred in pursuit of human missions to Mars, public commitment over several decades will be critical to mission success. Because the public is a primary stakeholder, a traditional outreach program is insufficient. The term "public engagement" is important philosophically; it differs from "outreach" in that it is, by nature, two-directional and implies that it is no longer just about reaching out. Active participation and communication back into the program is extremely important to enable the public to take part authentically in discovery and exploration. While many technical decisions must be made by mission experts who will carefully consider and select among options that best support safe arrival, this decision process can be made more transparent. Special opportunities must be created where public input can be included without increasing mission risk, including decisions related to the type of "public engagement payloads" that are of interest, and a public role in their selection (e.g., the most popular request currently is for a microphone so people can hear how Mars sounds). Sharing the adventure with video feeds and interactive sessions with astronauts from the moon and Mars can only happen if public engagement is considered early in mission design and among the principle requirements for mission success (e.g., decisions related to increasing bandwidth and reserving mass, power, and space for additional payloads that may not be strictly scientific or life-supporting).

This kind of active public participation at all stages will be a radical change in the way space exploration is conducted, and, in some ways, is as bold a vision as the venture to go beyond Earth. It will take a dedicated public engagement program to create strategic pathways for enabling increasingly sophisticated and informed public participation in the human exploration of Mars. NASA has long been a civilian space agency, but it now has the opportunity to become increasingly a citizen space agency, a modern transformation worthy of epic exploration in the 21st century and of this momentous era in human history.

8.1 Guiding Principles for Public Engagement

To create a rich environment for public engagement, five guiding principles that are related to the human exploration of Mars include: story, participation, connectivity, inclusion, and transparency. These are addressed below.

a. *Story*: Story has, since ancient times, been central to human understanding of the world around us and to communicating knowledge from one generation to the next. What is already inherent in sending humans to Mars are the chapters and story arc of exploration.

b. *Participation*: For a strong public engagement program, good concepts that are imbedded within mission plans are those that allow the public to follow along and experience the adventure as it happens. The best concepts, however, are those that allow the public to participate actively in the process of discovery.

c. *Connectivity*: While they should be tailored to meet the needs of individual audiences, public-engagement activities should also be designed to encourage partnerships that connect one group to another – industry to schools, museums to universities, media to civic organizations, and all manner of networks – to provide the richest interactions, the sharing of knowledge, enhanced technical literacy, and a connection to others.

d. *Inclusion*: This journey beyond our home planet belongs to everyone. Whether it is accomplished by one nation or several, the lasting significance of this endeavor is of global scope. Greater opportunities for encouraging those who have been traditionally under-represented in science and engineering fields should remain a key in all public-engagement programming to ensure that NASA can attract and retain human talent and ingenuity from across the nation, among citizens of all backgrounds. Listening to the viewpoints and ways of knowing of indigenous peoples worldwide can also help guide a thoughtful approach to the design of exploration that supports survivability and sustainability on this world, as well as on others. The moon and, to a lesser extent, Mars are considered sacred bodies to some cultures, so this kind of dialog is vital to helping avoid unnecessary misunderstandings and to designing a program that is respectful of all traditions. Regardless of specific cultural traditions, the moon is iconic to all of humanity, and, along with the sun, is one of the first bodies pointed out to young children the world over. Because the moon is part of everyone's nighttime experience, people care about the it and are beginning to care about Mars as familiarity with it and its potential as a human destination increases. These sentiments cannot be ignored for bodies that might be considered, from the public perspective, part of the global commons no matter which nations are the first to visit (or in the case of the moon, revisit) them.

e. *Transparency*: For an agency to be transparent, it must capture the public's trust by being entirely forthright in all its decisions and actions. By encouraging active participation on the part of the public from the inception of the program, NASA can achieve transparency in the public arena.

8.2 Public Engagement Strategy

While beyond the scope of this study, a detailed plan for public engagement must be created that is based on formative analyses of the ways in which the national and global public would like to participate in the adventure. Without this public input, it is premature to select definitively an action plan for public engagement. At the same time, what likely binds Earth, moon, and Mars exploration is a central organizing theme that is both immediate and compelling in human terms: **survivability and sustainability on any of these worlds**.

In fact, to say that the human exploration of Mars is a civilization endeavor means that there can no longer be a strict separation between the majority of citizens and "rocket scientists" who specialize in space careers. At the same time, public input cannot be random and whimsical; the emphasis has to be on informed public participation in a manner that complements and enhances NASA goals. To enable citizens to gain the expertise that is necessary to become full-fledged members of an increasingly spacefaring society, a progressive pathway for participative learning experiences should be created. For cohesion, these progressive learning experiences, which are designed to build knowledgeable public participation, will center on three major strands of public engagement that are related to survivability and sustainability on the Earth, moon, and Mars: science, technology, and society. These can be defined as follows:

- *Science*: Acquiring place-based knowledge through imaging and data analysis
- *Technology*: Developing a human-robotic partnership through innovations and inventions
- *Society*: Building a shared human experience through the arts, humanities, and social sciences

Public engagement activities in each of the three topical strands of science, technology, and society, acting in partnership with complementary activities in formal education, will deepen and expand in concert with missions that are pursued during the technical phases of Mars exploration: reconnaissance, intensive investigation, sampling, human precursors (moon), human exploration (Mars), and a sustained human presence (figure 8-1). Public engagement pathways will progressively develop citizen capabilities in each of the three topical strands to the extent that participation in the human exploration of Mars is no longer remote and, instead, is able to be imbedded within the context of people's communities and lives.

	Central Organizing Theme: Survival & Sustainability for the Earth, Moon, and Mars					
	Pathways to a Public Engagement Outcome By Mars Exploration Phase					Public Engagement Outcome
	Reconnaissance	Intensive Investigation	Sampling	Human Precursor (Moon)	Human Exploration (Mars) / Sustained Human Presence	
	Heritage Public Engagement	Expanded Heritage	Potential New Efforts			
SCIENCE: Acquiring Place-based Knowledge through Imaging & Data Analysis	Earth/Mars Comparisons Rock Around World; Mars Student Imaging & Analysis	Public "Earthonaut" Investigations in Local Environments; Earth, Moon, & Mars Data-mining Credentialing	Earthonaut Local Soil Comparisons to Mars Soil Simulants "Returned"; Earth, Moon, & Mars Data-mining	Citizen-Directed Investigations on the Moon	Citizen-Directed Investigations on Mars	Citizen Scientists gaining new knowledge...
TECHNOLOGY: Developing a Human-Robotic Partnership through Inventions & Innovations	FIRST, LEGO, BEST, Weather Balloons	Public Design Challenges: Survivor Rover; Habitat Technologies for Humanity	Public Design Challenges: Human Helper Robots; "Invention Convention" Design Challenges for Earth Habitats (food, transpo, energy etc.)	Citizen-Selected Public Engagement Payloads for the Moon; Bioplex Dome Testbed for Citizen/Industry Research related to Outpost Living on the Moon & Mars	Citizen-Selected Public Engagement Payloads for Mars; Habitat Technologies Integrated on the Earth, Moon, & Mars	...and utilizing technology for sustainable living & personal exploration...
SOCIETY: Building a Shared Human Experience through the Arts, Humanities, & Social Sciences	Citizen Think Tanks; Imagine Mars Project IMAX/Nova programs; Sundial Messages, Send Name to Mars	Citizen Councils; Virtual Field Trips & Simulations for Outpost Living; Time Capsules for Mars Created	Citizen Council Input for Sample Returns; Reality Earth Earth Classroom	Citizen Council Input Integrated for Moon; Reality Moon Lunar Classroom	Citizen Council Input Integrated for Mars; Reality Mars Martian Classroom; Time Capsules for Mars Opened	...as part of a Space-faring Society.

Figure 8-1. Sample public engagement pathways.

APPENDIX A: MEMBERSHIP

Agency Steering Group
Juan Alonso	Aeronautics Research Mission Directorate
Douglas R. Cooke[†]	Exploration Systems Mission Directorate
Dan Durda	Science Mission Directorate
Mike Hawes	Space Operations Mission Directorate
David Radzanowski	Space Operations Mission Directorate
David R. Liskowsky	Chief Health and Medical Office
Doug McCuistion[†]	Science Mission Directorate
Lisa Porter	Aeronautics Research Mission Directorate
Geoffery Yoder	Exploration Systems Mission Directorate

Mars Architecture Study Strategy Group
Dave Beaty[†]	Jet Propulsion Laboratory
Vicki Crisp	NASA Headquarters
Bret Drake[†]	Johnson Space Center
Scott Goodwinn	NASA Headquarters
George Tahu[†]	NASA Headquarters
Jeff Volosin	NASA Headquarters

Integration Team
David Bearden	The Aerospace Corporation
Robert Bitten	The Aerospace Corporation
Cassie Conley	NASA Headquarters
Bret Drake[†]	Johnson Space Center
John Elliott	Jet Propulsion Laboratory
Joe Fragola	Valador, Inc.
Peter Gage	Valador, Inc.
Arlene Moore	NASA Headquarters
Hoppy Price[†]	Jet Propulsion Laboratory
Blake Putney	Valador, Inc.
Torrey Radcliffe	The Aerospace Corporation

Entry, Descent, and Landing Team
James Arnold	Ames Research Center
Deepak Bose	Ames Research Center
Jeff Bowles	Ames Research Center
Lee Bryant	Johnson Space Center
Chris Cerimele	Johnson Space Center
Neil Cheatwood	Langley Research Center
Scott Coughlin	Johnson Space Center
Juan Cruz	Langley Research Center
Donald Durston	Ames Research Center
Alicia Dwyer Cianciolo	Langley Research Center
Walt Engelund[†]	Langley Research Center
Chirold Epp	Johnson Space Center
David Kinney	Ames Research Center
Bernnie Laub	Ames Research Center
Robert Manning[†]	Jet Propulsion Laboratory
Robert Mueller	Kennedy Space Center
Richard Powell	Langley Research Center
Mike Tauber	Ames Research Center
Mike Tigges	Johnson Space Center
Ethiraj Venkatapathy	Ames Research Center
Carlos Westhelle	Johnson Space Center
Henry Wright	Langley Research Center
Michael Wright	Ames Research Center
Seokkwan Yoon	Ames Research Center

Precursors Team
Robert Easter	Jet Propulsion Laboratory
Andy Gonzales	Ames Research Center
Frank Jordan[†]	Jet Propulsion Laboratory
Larry Lemke[†]	Ames Research Center
Richard Mattingly	Jet Propulsion Laboratory
Carol Stoker	Ames Research Center

Human Health and Performance Team
Mary Van Baalen	Johnson Space Center
Steve R. Blattnig	Langley Research Center
John Charles[†]	Johnson Space Center
Martha S. Clowdsley	Langley Research Center
Francis A. Cucinotta	Johnson Space Center
Steve B. Guetersloh	Johnson Space Center
Jeff Jones	Johnson Space Center
Kathy Laurini	Johnson Space Center
Robert D. Rutledge	Johnson Space Center
Eddie J. Semones	Johnson Space Center
Lisa C. Simonsen	Langley Research Center
Frank M. Sulzman	Johnson Space Center
E. Neal Zapp	Johnson Space Center

Human Exploration of Mars Science Analysis Group
Ariel D. Anbar	Arizona State University
Mary Sue Bell	Johnson Space Center
R. Todd Clancy	Space Science Inst.
Charles S. Cockell	Open University, UK
Jack E. Connerney	Goddard Space Flight Center
Gregory Delory	UC Berkley
Jay T. Dickson	Brown University
Peter T. Doran	University of Illinois at Chicago
Rick C. Elphic	Ames Research Center
Dean B. Eppler	Johnson Space Center
David C. Fernandez-Remolar	Center De Astrobiologica INTA
Gian Gabriele	University d'Annunzio, Italy
James B. Garvin[†]	Goddard Space Flight Center
John Gruener	Johnson Space Center
James W. Head III	Brown University
Jennifer Heldmann	Ames Research Center
Mark Helper	University of Texas
Victoria Hipkin	Canadian Space Agency
Melissa D. Lane	Planetary Science Inst.
Joel S. Levine[†]	Langley Research Center
Joseph Levy	Brown University
Jeff Moersch	University of Tennessee
Lewis Peach	USRA
Francois Poulet	CNRS France
James W. Rice	Arizona State University
Kelly J. Snook	NASA Headquarters
Steven W. Squyres	Cornell University
James R. Zimbelman	Smithsonian Institution

Flight and Surface Systems Team

Fernando Abilleira	Jet Propulsion Laboratory
James Arnold	Ames Research Center
Bob Bagdigian	Marshall Space Flight Center
Stan Borowski	Glenn Research Center
Bob Cataldo	Glenn Research Center
Tracie Crane	Marshall Space Flight Center
Chris Culbert	Johnson Space Center
George Culver	Johnson Space Center
Monica Doyle	Johnson Space Center
Chad Edwards	Jet Propulsion Laboratory
Harvey Feingold	Johnson Space Center
Stuart Feldman	Marshall Space Flight Center
J Garcia	Ames Research Center
Mike Gernhardt	Johnson Space Center
Brand Griffin	Marshall Space Flight Center
Jerry Hill	Langley Research Center
Steve Hoffman[†]	Johnson Space Center
David Kinney	Ames Research Center
Ronald Leung	Goddard Space Flight Center
David Manzella	Glenn Research Center
Robert Menrad	Goddard Space Flight Center
Phil Metzger	Kennedy Space Center
Kenneth B. Morris	Marshall Space Flight Center
Jack Mulqueen	Marshall Space Flight Center
Gary Noreen	Jet Propulsion Laboratory
R.M. Ondler	Johnson Space Center
David Plachta	Glenn Research Center
Lloyd Purves	Goddard Space Flight Center
Shawn Quinn	Kennedy Space Center
Monserrate C. Roman	Marshall Space Flight Center
Carol Russo	Ames Research Center
Mike Sander	Jet Propulsion Laboratory
Jerry Sanders	Johnson Space Center
Paul Schenker	Jet Propulsion Laboratory
Jim Schier	NASA Headquarters
Biren Shah	Jet Propulsion Laboratory
Phil Sumrall	Marshall Space Flight Center
Larry Toups	Johnson Space Center
Huy Tran	Ames Research Center
Terry Tri	Johnson Space Center
Abhi Tripathi	Johnson Space Center
Timothy Tyburski	Glenn Research Center
Ethiraj Venkatapathy	Ames Research Center
David Waits	Marshall Space Flight Center
Kevin Watson	Johnson Space Center

Goal IV+ Team

Richard Alena	Ames Research Center
Geoffery Briggs	Ames Research Center
Debra Bubenheim	Ames Research Center
I Joseph Burt	Goddard Space Flight Center
Bill Clancey	Ames Research Center
Doug Craig[†]	NASA Headquarters
Brian Glass	Ames Research Center
Andrew Gonzales	Ames Research Center
Arnold James	Ames Research Center
Joshi Jitendra	NASA Headquarters
Harry Jones	Ames Research Center
John Karcz	Ames Research Center
Russel Kerschman	Ames Research Center
Jen Keyes	Langley Research Center
Mark Kliss	Ames Research Center
Rob Landis	Johnson Space Center
Bill Larson	Kennedy Space Center
Tony Lavoie	Marshall Space Flight Center
Pascal Lee	Ames Research Center
Darlene Lim	Ames Research Center
Bernadette Luna	Ames Research Center
Robert W. McCann	Ames Research Center
Chris Moore	NASA Headquarters
Louis Ostrach	NASA Headquarters
Pat Troutman	Langley Research Center
Bruce Webbon	Ames Research Center
Jhony Zavaleta	Ames Research Center

Heilophysics/Astrophysics Team

Mario Acuna	Goddard Space Flight Center
Thomas Cravens	University of Kansas
Dick Fisher	NASA Headquarters
Jeff Forbes	University of Colorado
Barbara Giles	NASA Headquarters
Don Hassler	Southwest Research Institute
Tim Killeen	National Center of Atmospheric Research
Janet Luhmann	UC Berkley
Paul Mahaffy	Goddard Space Flight Center
Chris McKay	Ames Research Center
Tom Moore	Goddard Space Flight Center
Neil Murphy	Jet Propulsion Laboratory
Larry Paxton	Applied Physics Laboratory
Arik Posner	NASA Headquarters
Michael Salamon	NASA Headquarters
Jim Slavin	Goddard Space Flight Center

[†] Denotes team leads, group chair or co-chair

APPENDIX B: REFERENCES

1. MEPAG HEM-SAG (2008). Planning for the Scientific Exploration of Mars by Humans. Unpublished white paper (J. B. Garvin and J. S. Levine, Editors) posted March 2008 by the Mars Exploration Program Analysis Group (MEPAG) Human Exploration of Mars Science Analysis Group (HEM-SAG) at http://mepag.jpl.nasa.gpv/reports/index.html.
2. Jakosky, B. M. and R. J. Phillips, 2001: Mars' Volatile and Climate History, *Nature* **142**, 237.
3. MEPAG (2006), Mars Scientific Goals, Objectives, Investigations, and Priorities: 2006, J. Grant, ed., 31 p. white paper posted February, 2006 by the Mars Exploration Program Analysis Group (MEPAG) at http://mepag.jpl.nasa.gov/reports/index.html.
4. Gault, D.E. (1970) Saturation and equilibrium conditions for impact cratering on the Lunar surface: Criteria and implications. *Radio Science* **5**, 273-291.
5. Hartmann, W.K. (1972) Paleocratering of the Moon: Review of post-Apollo data. *Astrophysics Space Science* **17**, 48-64.
6. Barlow, N.G. (1988) Crater size/frequency distributions and a revised relative Martian chronology. *Icarus*, 74, 285-305.
7. Strom,R.G., Croft, S.K., Barlow, N.G. (1992) The Martian imact cratering record. In *Mars* (H.H.Kieffer, B.M.Jakosky, C.W.Snyder, and M.S.Matthews eds.), University of Arizona Press, 384-423.
8. Neukum, G., Ivanov, B.A., Hartmann, W.K. (2001) Cratering records in the inner solar systemin relation to the Lunar reference system. *Space Science Reviews* **96**, 1-4, 55-86.
9. Nyquist, L.E., Bogard, D.D., Shih, C.-Y., Greshake, A, Stoffler, D., Eugster, O. (2001) Ages and geologic histories of Martian meteorites. *Space Science Reviews* **96**, 1-4,105-164.
10. National Research Council (2006) Assessment of NASA's Mars Architecture 2007-2016, Committee to Review the Next Decade Mars Architecture, Space Studies Board, Division on Engineering and Physical Sciences. The National Academies Press, Washington, D.C.
11. Cantor, B. A., P. B. James, M. Caplinger, and M. J. Wolff (2001) Martian dust storms: 1999 Mars Orbiter Camera observations, Journal of Geophysical Ressearch 106, E10, 23653-23688.
12. Nair, H., M. Allen, A. D. Anbar, Y. L. Yung, and R. T Clancy (1994) A photochemical model of the martian atmosphere, *Icarus* **111**, 124-150.
13. Kahn, R. (1984) The spatial and seasonal distribution of Martian clouds and some meteorological implications, Journal of Geophysical Research **89**, 6671-6688.
14. Jakosky, B. M. and C. B. Farmer (1982) The seasonal and global behavior of water vapor in the Mars atmosphere-Complete results of the Viking atmospheric water detector experiment, Journal of Geophysical Research **87**, 2999-3019.
15. Clancy, R. T., A. W. Grossman, M. J. Wolff, P. B. James, D. J. Rudy, Y. N. Billawala, B. J. Sandor, S. W. Lee, and D. O. Muhleman (1996) Water vapor saturation at low altitudes around aphelion: A key to Mars climate?, *Icarus 122*, 36–62.
16. Levine, J. S. (Editor) (1985) The Photochemistry of Atmospheres: Earth, The Other Planets, and Comets, Academic Press, Inc., Orlando, 518 p.
17. Yung. Y. L. and W. B. DeMore, 1999: Photochemistry of Planetary Atmospheres, Oxford University Press, New York, 456 p.
18. Formisano, V, S. Atreya, T. Encrenaz, N. Ignatiev, and M. Giuranna (2004) Methane on Mars, *Science* **306**, 5702, 1758-1761.
19. Krasnopolsky, V. A., J. P. Maillard, and T. C. Owen (2004) Detection of methane in the martian atmosphere: evidence of life?, *Icarus* **172**, 537-547.
20. National Research Council (2007) An Astrobiology Strategy for the Exploration of Mars. The National Academies Press, Washington, DC, 118 pp.
21. MEPAG SR-SAG (Special Regions Science Analysis Group) (2006), Findings of the Mars Special Regions Science Analysis Group, Astrobiology 6, 677-732. The document can also be accessed at http://mepag.jpl.nasa.gov/reports/index.html.

www.ingramcontent.com/pod-product-compliance
Lightning Source LLC
Chambersburg PA
CBHW081731170526

45167CB00009B/3784